John Westley Wright

A Text Book of Ophthalmology

John Westley Wright

A Text Book of Ophthalmology

ISBN/EAN: 9783337114077

Printed in Europe, USA, Canada, Australia, Japan

Cover: Foto ©berggeist007 / pixelio.de

More available books at **www.hansebooks.com**

A TEXT BOOK

OF

OPHTHALMOLOGY

BY

JOHN W. WRIGHT, A. M., M. D.

Professor of Ophthalmology and Clinical Ophthalmology in the Ohio
Medical University; Ophthalmologist to the Protestant
and University Hospitals, Columbus, Ohio.

COLUMBUS:
J. L. Trauger, Printer and Publisher
1896

PREFACE.

THE object of this treatise is to provide the medical student with a systematic text in the primary principles of ophthalmology, such as will be a reliable assistance to him in his pursuits of knowledge as a student of medicine.

No apology can be necessary for the issuance of a work of this kind, when it is known that there is not a treatise that is particularly designed for the student in his class work in college, most works being intended for the skilled oculist.

The arrangement of the subjects has been made in keeping with the needs of the student: first the anatomy and physiology of the eye, then the diseases and their treatment in such natural order, that the subject treated is designed to prepare him as well as possible at each recitation for that which follows.

While this treatise is especially intended for the student in his class work at college, it is no less adaptable for him as a general practitioner, for what is necessary for the student to know and understand thoroughly, in reference to a knowledge of the anatomy, diagnosis and treatment of an affection of the eye, in the most concise and practical form, is equally suited to the general practitioner.

Another important feature of this work is the glossary, or rather a limited dictionary of not only all the terms used in this treatise, but all terms relating to the eye so far as it has been possible to gather them. This I conceive will serve to save time in searching for the meaning of a word and often prevent the student from passing it entirely, for if his medical dictionary should not happen to be at hand, the word is often passed and perhaps not referred to again.

The pronunciation of the word is also a feature not to be neglected by the student, for whatever pronunciation he has been allowed to use at school, be it right or wrong, is generally carried throughout life; for this reason it has been the effort of the author to obtain the latest up-to-date pronunciation of all words used in this text. This part of the work has been under the supervision of Prof. G. M. Waters, to whom I am under many obligations.

As the plan of recitation in medical schools is fast superseding that of the lecture, it has been my endeavor to so arrange the matter that it may be especially adapted to that method.

The subjects, therefore, are conveniently arranged in sections of about one recitation each, and the paragraphs are so condensed that the teacher can, at a glance, intelligently frame his question.

I am much indebted to my late assistant, Dr. D. L. Cowden, and my son, C. C. Wright, for their kind assistance in compiling the glossary. This has been an arduous work, the completeness of which attests the painstaking endeavor to make it as precise, and at the same time, as explicit as possible.

Many of the definitions in the glossary have been taken word for word from Duane and Thomas.

As it has been my desire to obtain the most explicit definitions, and as I have found those which I have taken from the above authors all that could be desired, I herein acknowledge my indebtedness to them.

During the preparation of these pages, the following works have been freely consulted, and any matter which was considered as particularly advantageous to the student, has been utilized and arranged to conform with the conception of this work.

Diseases of the Eye. *DeSchweinitz.*

Text-Book of Ophthalmology. *Fuchs.*

Diseases of the Eye. *MacNamara.*

Swanzy on the Eye.

Human Physiology. *Flint.*

Gray's Anatomy.

Leidy's Anatomy.

Ophthalmology and Ophthalmoscopy. *Schmidt-Rimpler, by Roosa.*

Hare's Therapeutics.

Student's Aid in Ophthalmology. *Walker.*

Diseases of the Eye. *Noyes.*

Diseases of the Eye. *Meyer.*

It now remains for me to particularly acknowledge the kind assistance of my friend, Mr. Fred J. Heer, who, together with the employes of the Lutheran Book Concern, has had charge of the mechanical work. It has been their constant effort to bring this part to the highest possible standard.

JOHN W. WRIGHT.

141 East Long Street, Columbus, O.

LIST OF ILLUSTRATIONS.

CONTENTS.

ERRATA.

Page 3, *for* "foraminea" *read* "foramina."
" 3, *for* "infraorbita" *read* "infraorbital."
" 4, *for* "fossae" *read* "fossa."
" 7, *for* "conjunctiva" *read* "conjunctivae."
" 10, 41, 43 and 76, *for* "palpebra" *read* "palprebae."
" 13, *for* "Descement's" *read* "Descemet's."
" 17, *for* "tendenously" *read* "tendinously."
" 23, *for* "are" *read* "is."
" 33, *for* "retina" *read* "retinae."
" 41, *for* "palpebra" *read* "palpebrarum."
" 52, *for* "preferance" *read* "preference."
" 54, *for* "subaceous" *read* "sebaceous."
" 61, *for* "dacyrocystitis" *read* "dacryocystitis."
" 62, *for* "abcess" *read* "abscess."
" 96, *for* "muce" *read* "muco."
" 117, *for* "pterygium" *read* "pterygium."
" 157, *for* "Decemet's" *read* "Descemet's."
" 165, *for* "tract" *read* "tracts."
" 169, *for* "Paracentisis" *read* "Paracentesis."
" 369, *for* "canaliculis" *read* "canaliculus."

CONTENTS.

CHAPTER I.

ANATOMY AND PHYSIOLOGY OF THE EYE.

CHAPTER II.

DISEASES OF THE ORBIT.

CHAPTER III.

DISEASES OF THE LACHRYMAL GLAND AND
LACHRYMAL APPARATUS.

CHAPTER IV.

INJURY AND DISEASES OF THE LIDS.

CHAPTER V.

DISEASES OF THE CONJUNCTIVA.

CHAPTER VI.

INJURIES AND DISEASES OF THE CORNEA.

CHAPTER VII.

INJURIES AND DISEASES OF THE SCLERA.

CHAPTER VIII.

DISEASES OF THE IRIS.

CHAPTER IX.

DISEASES OF THE CHOROID.

CHAPTER X.

DISEASES OF THE CILIARY BODY.

CHAPTER XI.

DISEASES OF THE RETINA.

CHAPTER XII.

DISEASES OF THE OPTIC NERVE.

CHAPTER XIII.

GLAUCOMA.

CHAPTER XIV.

DISEASES OF THE VITREOUS HUMOR.

CHAPTER XV.

CATARACT.

CHAPTER XVI.

SYMPATHETIC OPHTHALMIA — ENUCLEATION — ARTIFICIAL EYES.

CHAPTER XVII.

CHAPTER XVIII.

COLOR - BLINDNESS.

CHAPTER XIX.

EXTERNAL EXAMINATION OF THE EYE.

A TEXT BOOK

OF

OPHTHALMOLOGY

BY

JOHN W. WRIGHT.

OPHTHALMOLOGY.

CHAPTER I.

SECTION I.

THE EYE is the organ of vision.

VISION is the faculty of seeing. It is the process by which images of objects are made upon the retina and their impressions transferred to the brain.

OPHTHALMOLOGY is a treatise or a discourse upon the eye.

For convenience of study, Ophthalmology is divided into the ANATOMICAL, the PHYSIOLOGICAL, the PATHOLOGICAL and the THERAPEUTICAL.

ANATOMICAL OPHTHALMOLOGY treats of the anatomy of the eye and its appendages.

PHYSIOLOGICAL OPHTHALMOLOGY treats of the functions of the eye and its appendages.

PATHOLOGICAL OPHTHALMOLOGY treats of the diseases of the eye and its appendages.

THERAPEUTICAL OPHTHALMOLOGY treats of the remedial agents used in the treatment of the diseases of the eye and its appendages.

ANATOMICAL AND PHYSIOLOGICAL OPH-
THALMOLOGY.

The various diseases of the eye cannot be intelligently studied and appreciated without a knowledge of the parts which are in juxtaposition. Hence the anatomy of the ORBIT, the place where the eye rests, is of significant importance.

The ORBITS are the cavities which contain the eyes.

The orbits are four sided pyramidal bony cavities, with the bases in front and the apices behind.

Each orbit is composed of seven bones, viz.: superior maxillary, malar, frontal, palate, sphenoid, ethmoid, and lachrymal.

For convenience of study the bones of each orbit may be arranged as follows:

Bones of orbit.	Floor.	Orbital plate of superior maxillary. Portion of malar; orbital process of palate bone.
	Roof.	Orbital plate of frontal mainly; at apex small portion of sphenoid.
	Inner wall.	Mainly the ethmoid; lachrymal; sphenoid behind; nasal process of superior maxillary in front.
	Outer wall.	Sphenoid behind; malar in front.

The openings or foramina of the orbit may be considered as follows:

Orbital Foramina.	Optic.	Communicates with the cranial cavity.
	Sphenoidal.	Communicates with the cranial cavity.
	Spheno-maxillary.	Communicates with the temporal, spheno-maxillary, and zygomatic fossae
	Lachrymal.	Lachrymal fossa with the nose.
	Other foraminae.	The infra orbital, anterior and posterior ethmoid, and malar.

The *optic foramen* transmits the optic nerve and ophthalmic artery.

The *sphenoidal foramen* (fissure) transmits the third and fourth ophthalmic divisions of the fifth and sixth nerves, and the ophthalmic vein.

The *spheno-maxillary foramen* (fissure) transmits the infra-orbital vessels and nerves, and the ascending branches from the spheno-palatine ganglion, and superior maxillary nerve.

The *lachrymal foramen* (groove) is occupied by the lachrymal sac and duct, through which the tears are transmitted to the nose.

SMALLER FORAMINA.

Infra orbital foramen (groove) transmits infra orbita vessels and nerve.

The *malar foramen* (sometimes two) transmits facial nerves.

Anterior and *posterior ethmoidal foramina* transmit ethmoidal vessels and nasal nerve.

The orbit is lined by periosteum, continuous at the fissures and sutures with that of the bones of the face and the dura mater.

The periosteum lining the orbit forms a tendenous ring about the optic foramen; which gives origin to the ocular muscles.

The orbital periosteum is covered with a layer of connective tissue and fat, upon which the eye rests.

In the roof of the orbit are two depressions which require especial attention in the study of the eye:

1. The *forca trochlaris*, at the inner angle for the pulley of the superior oblique muscle.

2. The *fossae lachrymalis*, at the outer anterior angle for the lachrymal gland.

SECTION II.

THE LIDS.

The *lids* are the *appendages* of the eye.

The lid from without, inward, is composed of *integument, connective tissue, orbicularis palpebral muscle, tarsal cartilage, meibomian glands*, and *conjunctiva*.

The *integument*, which is continuous with that of the face, becomes continuous with the conjunctiva at the border of the lids.

The *cilia* are short stiff hairs placed in from

two to four rows in the integument at the anterior border of the lids.

The cilia protect the conjunctiva from dust and other foreign substances.

A *sebaceous follicle* is a sack containing a gland which opens at each side of the follicle of each cilium.

The functions of the cutaneous glands here are to moisten the cutaneous surface near the opening, and to soften the hairs.

These sebaceous glands are sometimes called Zeissian glands.

The *orbicularis palpebrarum muscle* surrounds the fissure between the lids, and is composed of pale thin fibres. It adheres closely to the integument anteriorly by the connective tissue, but posteriorly glides over the tarsal cartilages.

The opening formed by the edges of the lids is called the *palpebral fissure*.

The *tarsus* (plural,—tarsi. The term tarsal cartilage is an improper application.) A semilunar framework of condensed connective tissue, giving firmness and shape to either eye lid.

The tarsus in the upper lid is oval and is thickest at its anterior edge.

The tarsus in the lower lid is thinner and narrower than that of the upper lid and is of nearly uniform breadth throughout.

The tarsi are firmly held in position by fibrous tissue internally to the tendo oculi, externally to the malar bone, and above and below to the margin of the orbit, by the palpebral ligament.

The *Meibomian glands*, sometimes called paipebral glands, are sebaceous glands embodied in the under surface of each tarsus.

The Meibomian glands number from 25 to 30 in the upper lid, and from 20 to 25 in the lower. Each gland consists of an excretory duct, with caecal appendages arranged along its sides. These ducts open on the free borders of the posterior lip of the eye lid.

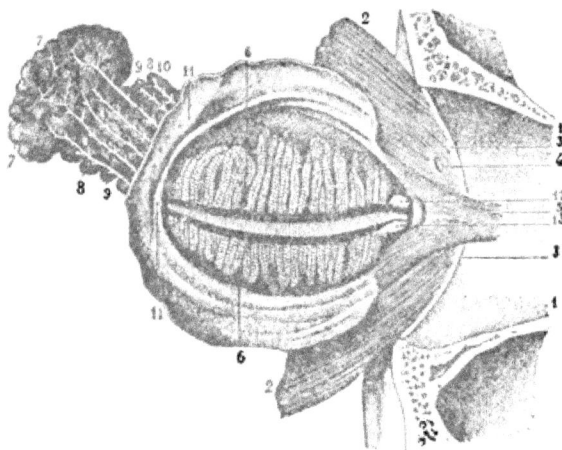

Lachrymal and Meibomian glands (Sappey).

1, 1, internal wall of the orbit; 2, 2, internal portion of the orbicularis palpebrarum; 3, 3, attachment of this muscle to the orbit; 4 orifice for the passage of the nasal artery; 5, muscle of Horner; 6, 6, posterior surface of the eyelids, with the Meibomian glands; 7, 7, 8, 8, 9, 9, 10, lachrymal gland and ducts; 11, openings of the lachrymal ducts.

FIGURE 1.

The Meibomian glands are simply large sebaceous glands whose function is to secrete the sebum, which lubricates the edges of the lids. This lubri-

cation of the edges of the lids prevents the overflow
of tears over their borders, and renders the palpebral
fissure water-tight.

THE CONJUNCTIVA. (MEMBRANA CONJUNCTIVA.)

The mucous membrane covering the inner surface of the eye lids and the outer surface of the eye
ball.

That part of the conjunctiva which covers the
inner surface of the lids is known as the *palpebral
conjunctiva;* the part covering the eye-ball, the *bulbar*
or *ocular conjunctiva.*

The *conjunctival sac* includes all that portion
anterior to the eye and posterior to the lids.

The *ocular conjunctiva* covers the anterior surface of the eye-ball, the anterior epithelium of the
cornea being a continuation of it.

Fornix conjunctiva is that portion of the conjunctiva at the point of reflection from the lid to
the globe.

The *caruncula lachrymalis* is a small rounded
projection at the inner angle of the eye, consisting
of a little island of cutaneous tissue bearing fine
hairs.

The palpebral conjunctiva is thickest and most
vascular, and is firmly adherent to the tarsus.

The fornix conjunctiva is thin and loose.

The scleral conjunctiva is thinner than the palpebral, and is loosely connected to the ball by episcleral tissue.

The corneal conjunctiva (almost entirely epithelial) is thinnest and very closely connected to the cornea.

The edges of the lids are lubricated by the secretion of the Meibomian glands.

The conjunctivæ are moistened by the secretion from the lachrymal gland.

THE LACHRYMAL APPARATUS.

The *lachrymal apparatus* is composed of the *lachrymal gland* and its *excretory* (lachrymal) *ducts*, the *lachrymal canaliculi*, the *lachrymal sac*, and *nasal duct*.

The *lachrymal gland* consists of two portions:— a large superior, and a small inferior portion, situated in a depression of the orbital plate, the fossa lachrymalis.

The *lachrymal ducts* convey the secretion of the lachrymal gland under the conjunctiva and open at the fornix conjunctivæ. They vary in number from seven to fourteen.

The *lachrymal canaliculi* are situated at the inner angle of the margin of each lid, and unite before reaching the lachrymal sac.

The *lachrymal sac* is the upper dilated portion of the passage which conveys the tears to the cavity of the nose.

The *nasal duct* extends from the lachrymal sac to the inferior meatus of the nose.

The *puncta lachrymalis* are the small openings or mouths of the canaliculi.

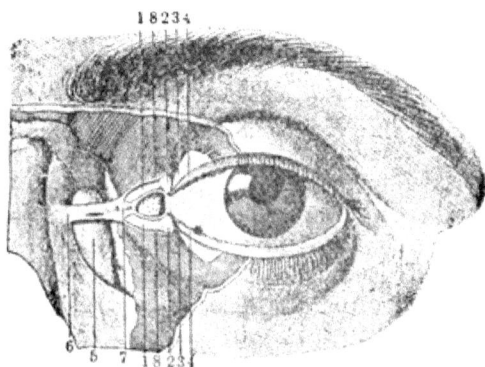

THE LEFT EYE, WITH A PORTION OF THE EYELIDS REMOVED, TO EXHIBIT THE LACHRYMAL CANALS AND SAC. 1, lachrymal canals; 2, commencement of these on the lachrymal papillae; 3, palpebral scutes; 4, edges of the eyelids; 5, lachrymal sac; 6, internal palpebral ligament; 7, its point of division in front of the lachrymal canals; 8, branches of the ligament giving attachment to the fibres of the palpebral orbicular muscle.

FIGURE 2. (After Sappey.)

The physiological functions of the lachrymal apparatus are the secretion of the tears by the lachrymal gland, their conveyance through the lachrymal (excretory) ducts to the fornix conjunctivae, whence the eye is moistened; the secretion is then taken up by the puncta and passes through the canaliculi to the lachrymal sac, thence through the nasal duct and into the nose.

The *lachrymal secretion* is a slightly alkaline solution, and serves to moisten the anterior portion of the eye.

By the frequent closure of the lids the secretion is conveyed from the lachrymal gland to the lachrymal sac.

The eyelids are opened by the contraction of the levator palpebræ superioris.

The eyelids are closed by the contraction of the orbicularis palpebræ and the relaxation of the levator palpebræ superioris.

The levator palpebra superioris is supplied by the third nerve.

The orbicularis palpebra is supplied by the facial nerve.

The point of union between the lids is called the palpebral commissure.

The *canthi* are the angles formed by the junction of the two lids.

The outer part of the fissure is known as the *outer canthus,* and that of the inner, as the *inner canthus.*

SECTION III.

THE EYE-BALL.

The eyeball is situated in the anterior part of the orbit to its outer side, and about equi-distant from its upper and lower walls. It rests upon a cushion of cellular tissue and receives protection in front by the eyelids.

The eye is spheroidal in form with the segment of a smaller sphere projecting from its anterior surface.

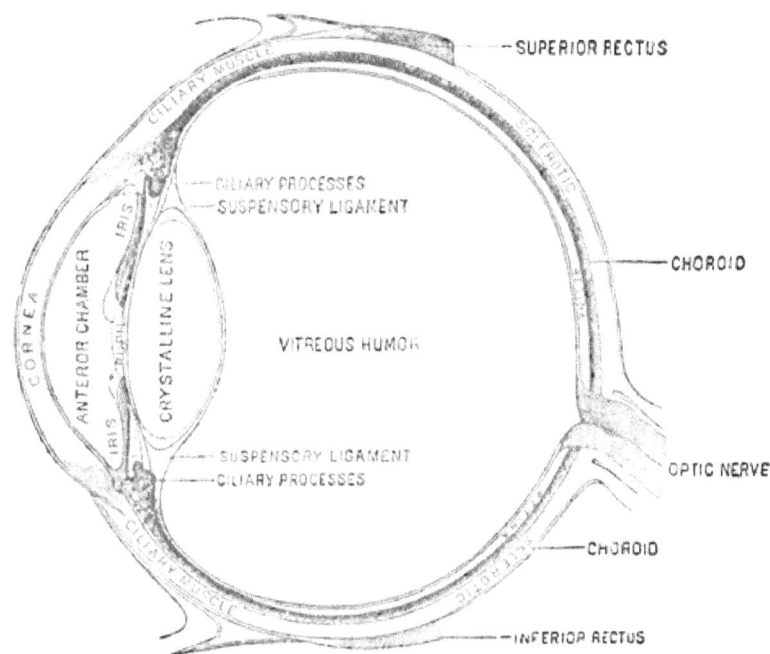

FIGURE 3.   After Flint.

The full size of the eye is not reached until about twenty years of age.

The *anterior pole* of the eye is the geometric center of the cornea.

The *posterior pole* or *fundus* of the eye is the geometric center of the back part of the globe.

The *optic axis* is an imaginary line from pole to pole. This line is also the antero-posterior diameter of the eye.

The *equatorial plane* is an imaginary plane through the center of the eye perpendicular to its axis, dividing the globe into the anterior and posterior hemispheres.

Meridional planes are imaginary planes coinciding with the axis.

The weight of the eye is about ninety-five grains.

The antero-posterior diameter of the eye is about ninety-seven hundredths of an inch. Its vertical diameter is about ninety-one hundredths of an inch, and its transverse diameter is about ninety-three hundreths of an inch.

The eyeball is composed of three tunics or coats, and from without inward are first, *sclerotica* and *cornea;* second, *tunica vasculosa;* third, the *retina.*

THE SCLEROTICA.

The *sclerotica* is the tough, white, outer fibrous coat covering the whole of the eye except the anterior portion occupied by the cornea.

The sclerotic coat serves to give shape to the eye and protects its more delicate interior.

The sclerotic is thickest at its posterior portion, and gradually becomes thinner as it approaches the cornea, where it again thickens at the sclero-corneal junction.

In infancy the sclerotic is very thin, and more transparent than in the adult, as is evidenced by the bluish tint due to the choroid which lines its inner surface.

Near the posterior axis of the eyeball is the *lamina cribrosa,* a sieve-like plate, consisting of many perforations of the sclerotic at this point, through

which the optic nerve, after being divided into many
bundles of nerve fibres, enters the eye.

The *canal of Schlemm* or circular venous sinus
is a congeries of blood vessels at the sclero corneal
junction, and parallel to the corneal border, but
entirely within the structure of the sclerotic.

The circular venous sinus is the outlet by which
the aqueous humor finds its way into the circulation.

The *cornea* is the transparent part of the external coat of the eyeball, and is situated in the
anterior portion of the eye.

The cornea projects beyond the curving surface
of the sclerotic, and is therefore the segment of a
smaller sphere than it (the sclerotic), and occupies
about one-sixth of the surface of the eyeball.

The cornea is about eleven mm. in its vertical
diameter, and twelve mm. in its horizontal.

The cornea is thicker at its rim than in the middle, being about one and one-tenth to one and two-
tenths mm. at the former and one mm. at the latter.

The cornea consists of five layers from without,
inwards, as follows:

(1st) *Epithelium;* (2nd) *anterior elastic lamina,*
(Bowman's membrane); (3rd) the *substantia propria,*
(cornea proper); (4th) *posterior elastic lamina,,* (Descement's membrane); and (5th) the *endothelium.*

The tissues of the cornea and sclerotic, at the
junction of the two pass imperceptibly into each
other.

The cornea has no blood vessels, except for a

narrow space about its rim of about one and five-tenths mm., at which place it is supplied from the conjunctiva, the vessels of which run between the epithelium and the anterior elastic membrane and turn upon themselves in capillary loops. These vessels are from the episcleral branches of the anterior ciliary arteries.

The cornea is abundantly supplied with nerves, derived from the ciliary nerves, which enter from the sclerotica as medullary fibres, and divide, after their entrance into the cornea, into transparent (non-medullated) branches.

The cornea receives its nourishment from a transparent liquid (the plasma) which circulates through the lymph spaces and channels of its substance.

TUNICA VASCULOSA.

The *tunica vasculosa* or uveal tract, the second tunic of the eye, consists of the *choroid, ciliary body* and *iris*. It lines the inner side of the sclerotica, and is perforated posteriorly by the optic nerve, and has a circular opening in front, the *pupil*.

The choroidal and ciliary portion of the tunica vasculosa is adhered more or less firmly to the entire inner surface of the sclerotica; the iritic portion, excepting the corneal margin, is free, and its surfaces are not adhered to any other part.

SECTION IV.

THE CHOROID.

The choroid lies between the sclerotica and retina, and extends from the optic nerve to the ciliary body. It is composed principally of blood vessels and dark brown pigment.

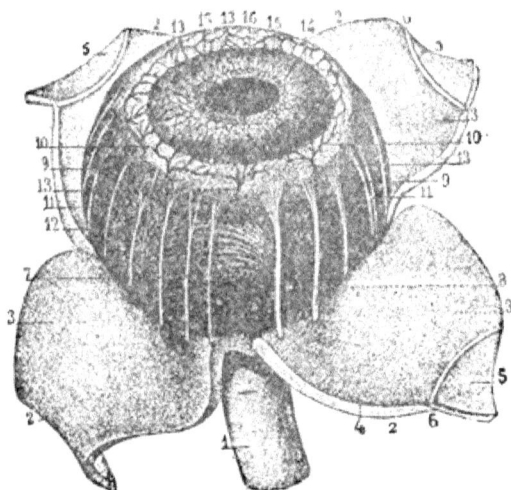

Choroid coat of the eye (Sappey).

1, optic nerve; 2, 2, 2, 2, 3, 3, 3, 4 sclerotic coat, divided and turned back to show the choroid; 5, 5, 5, 5, the cornea divided into four portions and turned back; 6, 6, canal of Schlemm; 7, external surface of the choroid, traversed by the ciliary nerves and one of the long ciliary arteries; 8, central vessel, into which open the vasa vorticosa; 9, 9, 10, 10, choroid zone; 11, 11, ciliary nerves; 12, long ciliary artery; 13, 13, 13, 13, anterior ciliary arteries; 14, iris; 15, 15, vascular circle of the iris; 16, pupil.

FIGURE 4.

Although the choroid is very thin, (from 1/300 to 1/150 of an inch in thickness) it is composed of four distinct layers from the sclerotic inward as follows:—*lamina fusca, tunica vasculosa propria, membrana chorio-capillaris, lamina elastica.*

1st. *Lamina fusca:* a dark thin layer, composed of loose connective tissue containing dark pigment cells, from which it gets its color and name (fuscus, dark). Its principal office seems to be to surround and protect the vessels and nerves passing forward to the ciliary body and iris.

There is a lymph space between it (the lamina fusca) and tunica vasculosa propria, the *supra-choroidal space.*

2nd. The *tunica vasculosa propria,* or the layer of larger choroidal arteries and veins.

These arteries are derived from the short posterior ciliaries (ten or twelve in number) which enter the sclerotica around the optic nerve, and proceeding forward expand in this membrane into many inosculating branches, some of which reach as far forward as the ciliary muscle, and there anastomose with the long and anterior ciliary arteries.

The veins of the tunica vasculosa propria, the vena vorticosa, are much larger than the arteries and are situated exterior to them.

3rd. The *membrana chorio-capillaris* is composed of the minute vessels which connect the arteries (arteria ciliaris posteriores breves) with the veins (vena vorticosa).

The functions of the chorio-capillaris are to convey the arterial blood to the venous system; to provide nutrition to the eye; to regulate intra-ocular tension; to supply nutrition and warmth to the outer layers of the retina, and in conjunction with the cil-

iary process, to supply nourishment to the vitreous.

4th. The *lamina elastica* or lining membrane,
a very thin hyaline membrane covering the inner
surface of the membrana chorio-capillaris, the layers of which are bound together by stroma, a fibrous
network in whose meshes are pigment-cells more or
less abundant in light eyes.

THE CILIARY BODY.

The *ciliary body* is that portion of the tunica
vasculosa which lies between the choroid and iris.
It is continuous with both and is about one-fourth
of an inch in width.

The ciliary body is composed of the *ciliary
muscles* and the *ciliary processes.*

THE CILIARY MUSCLES.

The *muscles* of the ciliary body are *radiating*
and *circular.* The *radiating fibres* of the ciliary
muscles are confined to the outer portion of the ciliary body, and arise tendenously from the sclero-
corneal junction, the posterior elastic lamina and
from the ligamentum pectinatum iridis, and are directed backwards and inwards where they terminate
in the choroid.

The inner portion of the ciliary muscle consists
of bundles of fibres which pursue a circular course
and are hence termed the circular fibres of the ciliary muscle.

The *circular fibres* of the ciliary muscle occupy

that portion of the ciliary body near its junction
with the iris.   The ciliary muscles are much developed in hyperopia.

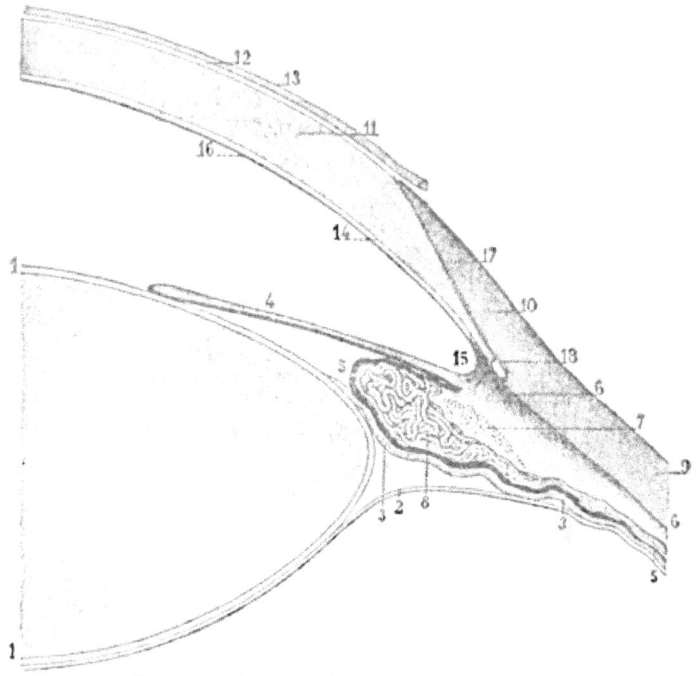

Ciliary muscle; magnified 10 diameters   Sappey).

1, 1, crystalline lens ; 2, hyaloid membrane ; 3, zone of Zinn ; 4, iris ; 5, 5, one of the ciliary processes ; 6, 6, radiating fibres of the ciliary muscle ; 7, section of the circular portion of the ciliary muscle ; 8, venous plexus of the ciliary process ; 9, 10,
sclerotic coat ; 11, 12, cornea ; 13, epithelial layer of the cornea ; 14, membrane of
Descemet ; 15, ligamentum iridis pectinatum ; 16, epithelium of the membrane of
Descemet ; 17, union of the sclerotic coat with the cornea ; 18, section of the canal
of Schlemm.

FIGURE 5.

In myopia they are less developed and are sometimes indeed almost entirely wanting.

The above conditions of the ciliary muscle are
very readily accounted for when we consider that
the circular fibres are the ones whose function it is
to provide for accommodation, and as accommoda-

tion is but little employed in myopia, they are not
properly developed, while the opposite condition exists in hyperopia, the circular fibres being constantly
called into action at every effort to see, become excessively developed.

THE CILIARY PROCESSES.

The *ciliary processes* are formed by folds of the
choroid at its anterior margin, covering the inner
surface of the ciliary body and forming a circle behind the iris around the margin of the lens. Each
one of these folds is called a process.

The ciliary processes number about seventy-
two, and are of variable size, the larger of which are
about one-tenth of an inch in length. The smaller
processes are superimposed between the larger ones,
but not in regular order.

The ciliary processes are deepest and thickest
at their fore part and gradually taper into the choroid behind. Their anterior extremities are rounded
and free, and are suspended in the aqueous humor
in a circle behind the outer border of the iris.

Within the folds of the ciliary processes are received corresponding folds of the thick membrane
called the *zone of Zinn*. This membrane is continuous with the anterior portion of the hyaloid membrane.

The ciliary body is richly supplied with nerves
from the long and short ciliary, and blood vessels
principally from the long ciliary arteries; hence that

part of the sclerotic immediately exterior to it is called the "danger zone," because any affection or injury of this structure is marked by extreme tenderness in this region of the eye.

The "*danger zone*" commences at the junction of the sclerotic and the cornea and extends backwards one-fourth of an inch.

SECTION V.

THE IRIS.

The *iris* is a membranous disk continuous with the ciliary body and choroid with a central opening the pupil. It is held in its position through the intervention of the ligamentum pectinatum iridis, by which its greater circumference is attached to the sclero-corneal junction. The iris also receives support from the lens, whose anterior surface is rested upon by the posterior surface of its pupillary border, thus pushing that part of the iris forward, causing it to be slightly convex in front.

The iris originates from the anterior surface of the ciliary body, being formed principally by the branches of the long ciliary arteries, which divide into two branches at the ciliary muscle on each side of the eye, and run in a direction concentric with the margin of the cornea, and uniting with the vessels of the opposite side form the *circulus arteriosus iridis major*.

From this extend the radial arteries to near the pupillary border, forming the *circulus arteriosus iridis minor.*

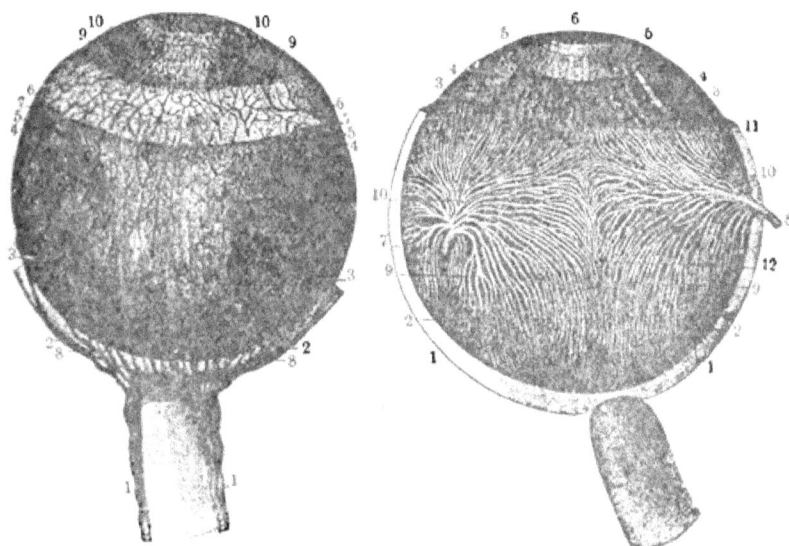

THE CHOROIDEA AND IRIS. 1, ciliary arteries situated at the sides of the optic nerve ; 2, the long ciliary arteries ; 3, the same after having pierced the sclerotica ; 4, 5, the main divisions of the same vessels ; 6, the ciliary muscle ; 7, the anterior ciliary arteries ; 8, the short ciliary arteries to the choroidea ; 9, the iris supplied by the long and anterior ciliary arteries ; 10, the pupil.

VEINS OF THE CHOROIDEA AND IRIS. 1, sclerotica ; 2, choroidea ; 3, ciliary muscle, of which a portion has been removed to exhibit the ciliary processes ; 4, 5, the iris ; 6, pupil ; 7, 8, trunks of the choroid veins ; 9, 10, vorticose vessels ; 11, their conjunction with the veins of the ciliary processes ; 12, anastomosis between the groups of vorticose vessels.

FIGURE 6. After Sappey.)

The vessels forming the iris are surrounded by a loose mesh-work of pigmented cells which fill the interspaces near the pupillary border; embedded in the stroma is the sphincter iridis, the contraction of which closes the pupil.

If the lens is absent, then the iris vibrates with the movement of the eye-ball. This condition of the iris is known as *iridodonesis*, and is caused by the loss of support which the iris receives from the lens.

The iris is divided into two zones: the *ciliary zone* and the *pupillary zone.*

The *ciliary zone* extends from the peripheral border of the iris to the circulus minor.

The *pupillary zone* extends from the circulus minor to the pupillary border of the iris.

These zones can best be distinguished in the blue eyes of children, where the pupillary margin is lined by a narrow black fringe.

They are also observed in those who are affected with cataract, where the white background of the opaque lens contrasts with the black fringe of the iris, thereby making them appear prominent.

The anterior surface of the iris is lined with a membrane continuous with the endothelium lining the posterior surface of the cornea. There are crypts or depressions in the iris which communicate with its tissues and place them (the tissue-spaces) into free communication with the anterior chamber of the eye. These crypts are not covered with endothelium.

The posterior surface of the iris is covered by the posterior lining membrane and the retinal pigment layer.

The diameter of the iris is about 12 mm. (one-half inch).

The pupil is subject to variations in size, ordinarily from 3 mm. to 5 mm. (⅛ to 1·5 inch).

The color of the iris depends upon the amount and tints of the particles contained in its pigment cells, and varies in different individuals.

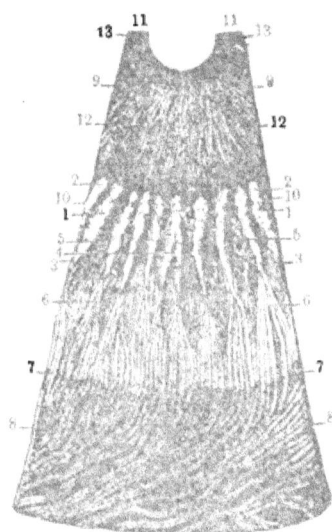

SEGMENT OF THE CHOROIDEA AND IRIS, SEEN ON ITS INNER SURFACE, magnified four diameters. 1, ciliary processes; 2, their free extremities behind the iris; 3, 4, commencement of the processes; 5, intervals of the processes; 6, veins of the ciliary processes; 7, posterior margin of the ciliary body; 8, choroidea with its veins; 9, iris; 10, its outer border; 11, the pupillary border; 12, radiating fibres of the iris; 13, circular fibres.

FIGURE 7. After Sappey.

The color of the irides as a rule are the same, but there are instances in which there are congenital differences although they may be perfectly healthy.

The iris by the action of the pupil in dilating and contracting, regulates the amount of light which enters the eye.

THE PUPIL.

The *pupil* is the aperture in the iris for the transmission of rays of light.

The pupil is normally eccentric, that is, slightly below and to the inner side. This condition is not noticeable without minute examination of the eye.

Corectopia is that condition in which the pupil is noticeably eccentric

Polycoria is that condition in which there is a multiplicity of pupils (that is, more than one).

A *natural pupil* is one which exists at birth.

An *artificial pupil* is one that has been made by art.

There is no standard size of the normal pupil, as so many contingencies exist even in health, which tend to change it; neither are there established means by which to record the exact measurement of the pupil.

The pupil in normal conditions varies in diameter from two mm. to six mm., the average diameter being about four mm.

The pupil is larger in children than in adults. In old age it becomes very small.

The pupil in the normal eye should be round and promptly contract and dilate to the effects of light and shade. It also contracts in the effort to accommodate and converge, that is to observe a near point, and dilates when the eye is adjusted for distant vision.

If one eye be shaded and the other exposed to a light, the pupil of the shaded eye acts in harmony with the other; this is termed the *consensual* action of the pupils.

The pupils of both eyes should be of uniform size under the same illumination.

Mydriasis is a morbid and excessive dilation of the pupil.

Mydriasis may be paralytic or spastic.

Paralytic mydriasis is produced by paralysis of the sphincter papilla or its supplying nerve.

Spastic mydriasis is produced by spasm of the dilator fibres of the iris, or stimulation of the sympathetic.

Myosis is an abnormal or excessive contraction of the pupil.

Myosis is paralytic or spastic.

Paralytic myosis is due to paralysis of the dilator fibres of the iris.

Spastic myosis is due to spasm of the sphincter pupilla.

There is also a contraction of the pupil known as *spinal myosis* which is regarded as a significant symptom in spinal disease, such as locomotor ataxia.

Myopes as a rule have large pupils and hyperopes small.

Hippus is a spontaneous, rapid and spasmodic dilation and contraction of the pupil. This condition is often observed in hysteria, mania, and other nervous affections. Its cause is obscure.

A *mydriatic* is an agent which produces dilatation of the pupil.

The principal mydriatics are atropine, duboisine, hyoscyamin and cocaine.

A *myotic* is an agent which causes contraction of the pupil.

The principal myotics are eserine, pilocarpine, and morphine.

Mydriasis or the dilatation of the pupil occurs in glaucoma, atrophy of the optic nerve, (although not constant) paralysis of the oculomotor nerves, paralysis of the nerves of the iris due to inflammation or increase of tension.

Myosis or contraction of the pupil occurs in hyperaemia or inflammation of the iris, in the early stage of all inflammatory affections of the brain and meninges, in diseases of the spinal chord.

Exclusion of the pupil is that condition in which the entire border of the iris is adherent to the capsule of the lens so that no fluid can pass through the pupillary space.

In *occlusion* of the pupil there is adhesion of the entire border of the iris to the lens capsule, but the pupillary space is filled with plastic lymph.

What is known as the Argyll Robertson pupil is that condition in which there is absence of reaction to light, although there is contraction of the already small pupil for accommodation and convergence.

The pupil in many cases reacts to light when there is no perception of light.

Light may be thrown into an eye whose pupil for some cause may not react, when its fellow will respond as promptly as if the light had been applied to the latter.

Irregularities in the dilation of a pupil under the effects of a mydriatic indicate the existence of *synechia.*

An inactive and dilated pupil does not always indicate a diseased condition of the retina.

SECTION VI.

THE RETINA.

The *retina* is the third or inner tunic of the eye.

The retina is a delicate transparent membrane formed from the fibres of the optic nerve, which are spread out in every direction from the entrance of the optic nerve to the ora serrata.

The retina is composed of two kinds of tissue, the *nervous tissue* and the *supporting tissue*, and these tissues assist in forming all the layers of the retina.

The retina is attached at two points only, at the optic nerve entrance, and at its anterior border, the ora serrata. The retina is not attached to the choroid, but simply lies on it.

In viewing the normal retina with the opthalmoscope, two points are particularly prominent, the optic disk and the macula lutea.

The *optic disk* is the head of the optic nerve, and marks its entrance into the globe.

The *macula lutea* is a thinning of the retina, and is marked by a central depression, the *fovea centralis.* It is about 1·6 mm. in diameter.

The microscopical examination of the retina demonstrates that it is composed, from within the globe outward, of the following layers:

1.   Membrana limitans interna.
2.   Fibrous layer.
3.   Vesicular layer.
4.   Inner molecular layer.
5.   Inner nuclear layer.
6.   Outer molecular layer.
7.   Outer nuclear layer.
8.   External limiting membrane.
9.   Jacob's membrane or rods and cones.
10.   Pigmentary layer.

1.   *Membrana limitans interna*, or the internal limiting membrane, is a thin transparent and imperfect membrane, said to be formed from the retinal connective tissue. It lies in contact with the hyaloid membrane of the vitreous humor. Its existence is considered doubtful by some histologists.

2.   The *fibrous layer* is formed by fibres of the optic nerve in their course to the ganglion cells.

3.   The *vesicular layer* is a single layer of large ganglion cells. Their structures are similar to those of the nerve centers.

4.   The *inner molecular layer* is made up of

fine fibres mingled with the processes of the ganglion cells.

5. The *inner nuclear layer* consists of two kinds of cellular elements, and two kinds of fibres. This layer is also known as the inner granular layer.

6. The *outer molecular layer* resembles the inner molecular layer, but contains branched stellate cells, supposed to be ganglion cells.

7. The *outer nuclear layer* consists of rod-granules and cone-granules and connective tissue elements.

8. The *external limiting membrane* is formed of retinal connective tissue, the terminal extremities of the fibres of Müller.

9. *Jacob's membrane* or *rods and cones* is composed of rods arranged perpendicularly to its surface, and cones with apices directed towards the choroid. This is the most important part of the retina, being the *percipient* layer.

10. The *pigmentary layer* consists of a single layer of hexagonal, nucleated cells. This layer was formerly considered part of the choroid.

The vascular supply of the retina is derived from the arteria centralis, with the exception of a slight anostomosis with the choroidal vessels at the optic disk.

The lymphatics of the retina exist around the vessels in the form of peri-vascular lymph-spaces.

The *rods* and *cones* are the percipient organs of the retina, and are connected by nerve-fibrils with the layer of nerve-fibres which convey the visual impulses through the optic nerve to the brain.

The point of most acute vision in the retina is the *macula lutea*, and for distinct vision the rays of light from an object must focus upon this point.

The objects of interest as seen in the retina of a normal eye through the ophthalmoscope, are *the optic disk, the macula lutea* and *the retinal vessels*.

A description of these parts will be fully given in the chapter on the ophthalmoscope.

THE OPTIC NERVE.

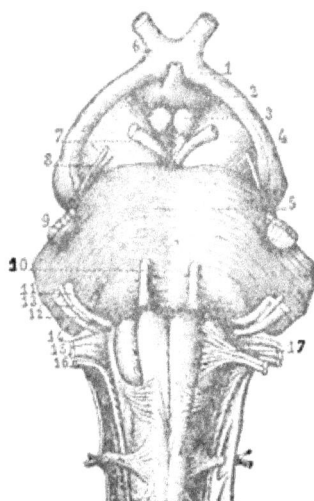

Optic tracts commissure and nerves Hirschfeld

1, infundibulum; *corpus cinereum;* 3, corpora albicantia; 4, cerebral peduncle; 5, pons Varolii; 6, *optic tracts and nerves, decussating at the commissure, or chiasm;* 7, motor oculi communis; 8, patheticus; 9, fifth nerve; 10, motor oculi externus; 11, facial nerve; 12, auditory nerve; 13, nerve of Wrisberg; 14, glosso-pharyngeal nerve; 15, pneumogastric; 16, spinal accessory; 17, sublingual nerve. *

FIGURE 8.

The *optic nerves* arise from the optic commissure, and pass forward and outward to the two optic foramina.

The optic nerves are formed at the commissure by a decussation of the fibres of the optic tracts.

The *optic tracts* are formed by two roots,—the *external* and the *internal*.

The *external root* takes its origin from three centers of gray matter, viz. the optic thalamus, the external geniculate body, and the anterior tubercles of the corpora quadrigemina.

The *internal root* arises from two centers of gray matter, viz. from the internal geniculate body and the posterior tubercles of the corpora quadrigemina.

These centers of gray matter which give origin to the optic tracts are connected to the cerebral cortex by a system of fibres known as the cortico optic radiating fasciculi and constitutes the most posterior part of the optic thalamus.

After the formation of the optic tract by the union of the external and internal roots, it passes forward along the posterior inferior surface of the optic thalamus, crosses the crus cerebri, traverses the side of the tuber cinerium, and in front of the infundibulum unites with the optic tract of the other side to form the *optic commissure*.

In the optic commissure the fibres of each optic tract undergo semi-decussation, that is, the outer fibres of each tract are continued into the nerve of the same side. The central fibres of each tract are continued into the optic nerve of the opposite side, decussating in the commissure and extend to the optic papilla or disk.

From its origin, the optic commissure, each optic nerve passes forward and outward to the optic foramen of its side.

The optic nerve is divided into three portions from within outward as follows:

1st. The *intra-cranial*, which part is included between the chiasm and the optic foramen.

2d. The *orbital*, which includes that portion between the optic foramen and the eyeball.

3d. The *intra-ocular*, which is found within the sclerotic and is the termination of the nerve.

The intra-cranial portion of the optic nerve is enclosed in a sheath derived from the arachnoid.

As the optic nerve passes through the optic foramen it receives an aditional covering from the dura mater which covers that portion of the nerve within the foramen.

As the optic nerve enters the orbit, this sheath from the dura mater subdivides into two layers, one portion of which is continuous with the periosteum of the orbit, and the other with the arachnoid surrounds the optic nerve as far as the sclerotic.

As the optic nerve enters the eyeball it becomes constricted in its diameter and looses its coverings from the arachnoid and dura mater, which become fused with the sclerotic coat.

The coarser interstitial connective tissue of the

nerve is intercepted by the lamina cribosa, and only the nerve-tubules and the blood vessels and some fine connective tissue elements are permitted to pass through into the interior of the globe.

Immediately anterior to the sclerotic foramen is the choroidal foramen, the margin of which is marked at parts of its entire circumference with black pigment.

The fibres of the optic nerve pass through these two foramina and immediately curve boldly round the margin of the choroidal foramen and spread out in all directions to form the anterior layer of the retina.

The optic nerve does not enter the eye at its posterior pole, but a little to the nasal side of it.

Just before the optic nerve enters the globe, it is perforated by a small artery, the arteria centralis retina, which traverses its interior in a canal of fibrous tissue, and supplies the inner surface of the retina. This artery is accompanied by corresponding veins.

The optic nerve is about four mm. (1·6 inch) in diameter and 28 mm. (1¦ inches) in length in its orbital portion.

The optic nerve is supplied by blood vessels from the ophthalmic artery.

The special and perhaps only function of the optic nerve is to convey impressions of sight to the cerebrum.

The optic nerve is not endowed with general sensibility and is therefore insensible to ordinary impressions.

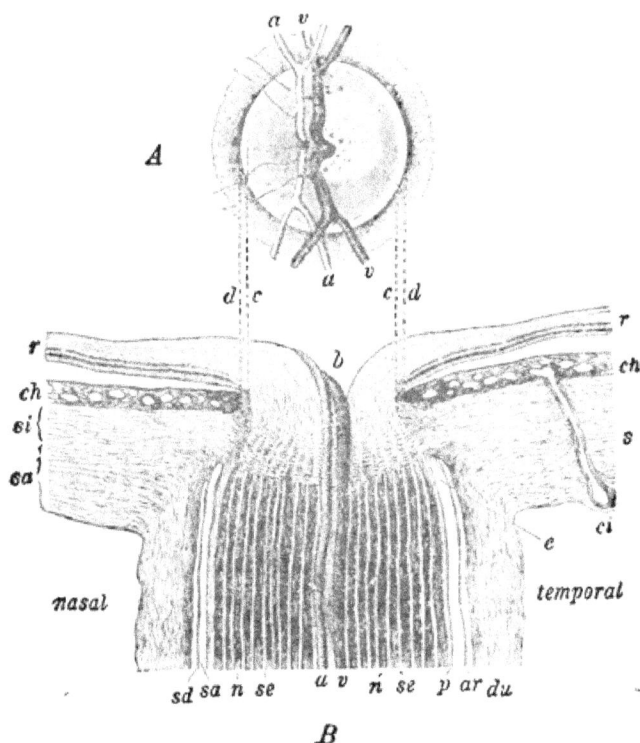

HEAD OF THE OPTIC NERVE.

A, OPHTHALMOSCOPIC VIEW. — Somewhat to the inner side of the center of the papilla the central artery arises from below, and to the temporal side of it rises the central vein. To the temporal side of the latter lies the small physiological excavation with the gray stippling of the lamina cribrosa. The papilla is encircled by the light scleral ring (between *c* and *d*) and the dark chorioidal ring at *d*.

B LONGITUDINAL SECTION THROUGH THE HEAD OF THE OPTIC NERVE. — Magnified 14 x 1. The trunk of the nerve up to the lamina cribrosa has a dark color because it consists of medullated nerve-fibers, *n*, which have been stained black by Weigert's method. The clear interspaces, *se* separating them, correspond to the septa composed of connective tissue. The nerve-trunk is enveloped by the sheath of pia mater, *p*, the arachnoid sheath, *av*, and the sheath of dura mater, *du*. There is a free interspace remaining between the sheaths, consisting of the subdural space, *sd*, and the subarachnoid space, *sa*. Both spaces have a blind ending in the sclera at *e*. The sheath of dura mater passes into the external layers, *sa*, of the sclera, the sheath of pia mater into the internal layers, *si*, which latter extend as the lamina cribrosa transversely across the course of the optic nerve. The nerve is represented in front of the lamina as of light color, because here it consists of non-medullated and hence transparent nerve-fibres. The optic nerve spreads out upon the retina *r*, in such a way that in its center there is produced a funnel-shaped depression, the vascular funnel, *b*, on whose inner wall the central artery, *a*, and the central vein, *v*, ascend. The chorioid, *ch*, shows a transverse section of its numerous blood-vessels, and toward the retina a dark line, the pigment epithelium is next the margin of the foramen for the optic nerve and corresponding to the situation of the chorioidal ring the chorioid is more darkly pigmented. *ci* is a posterior short ciliary artery which reaches the chorioid through the sclera. Between the edge of the chorioid, *d*, and the margin of the head of the optic nerve, *c*, there is a narrow interspace in which the sclera lies exposed, and which corresponds to the scleral ring visible by the ophthalmoscope.

FIGURE 9 A. & B. (Figure and description by Fuchs.)

OPTIC DISK.

The *papilla nervi optici* or *optic disk* is formed by the radiating fibres of the optic nerve immediately after their passage through the openings in the sclerotic and choroid.

The central artery of the retina emerges from the optic nerve rather to the inner side of the disk where it bifurcates, one branch of which passes vertically upwards and the other downwards to the retina.

The central vein accompanies the artery and is easily recognized by its darker color and larger size.

The optic disk, along its margin, is frequently marked by patches of black pigment, which represent spots of choroid. This is the margin of the choroidal foramina, therefore a black crescent lying against one side of the papilia or disk, or a black line entirely surrounding it, is a common physiological phenomenon.

The optic disk, therefore, is all that portion included within the margins of the choroidal foramen.

Internal to the choroidal foramen and at the edge of the disk is a whitish ring, the *sclerotic ring*, caused by the choroidal foramen being somewhat larger than the optic foramen, so that the edge of the sclerotic is seen as a white band through the transparent fibres. The sclerotic ring is not always complete and is more marked at the outer side of the disk owing to a thinning of the fibres of the optic nerve in that region.

The optic disk is usually round or oval in shape, although it may be very irregular in outline. It is about 1·5 mm. in its transverse diameter.

There is a depression near the center of the optic disk formed by the manner in which the nerve fibres change their direction in their passage to different portions of the retina. This depression is known as the *physiological excavation* or *cup.*

The optic disk is pierced in the center of the physiological excavation by the central retinal artery and vein.

The capillaries of the optic disk are supplied from three sources, viz. the short ciliary arteries of the choroid, the central artery of the retina, and the arterial twigs of the pial sheath.

The optic disk is really the head of the optic nerve, and is the only portion of the retina where the sense of vision is wanting, and is therefore called the *blind spot.*

SECTION VII.

THE MEDIA OF THE EYE.

The term *medium* in optics signifies any substance that will transmit light.

The *media* in ophthalmology are those parts of the eye which transmit the rays of light in the function of seeing. They are the *cornea*, the *aqueous humor*, *crystalline lens* and the *vitreous humor*.

The *cornea* is the clear transparent anterior portion of the sclerotic.

The *aqueous humor* is the colorless transparent liquid occupying that portion of the cavity of the eye lying between the cornea and crystalline lens. The space thus occupied is called the aqueous chamber.

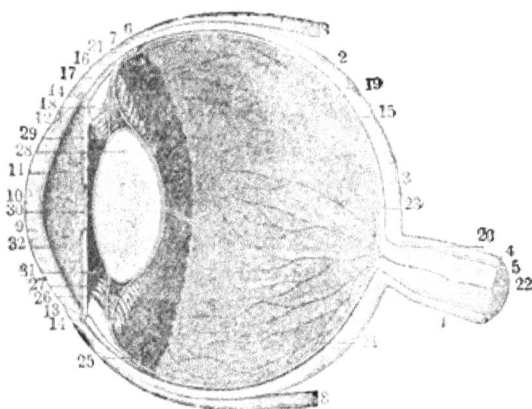

VERTICAL SECTION ANTERO-POSTERIORLY OF THE EYEBALL. 1, optic nerve; 2, sclerotica; 3, its posterior thicker portion; 4, sheath of the optic nerve continuous with the sclerotica; 5, the nerve within the sheath; 6, insertion of the recti muscles into the sclerotica; 7, 8, superior and inferior recti muscles; 9, cornea; 10, its conjunctival surface; 11, entocornea; 12, 13, bevelled edge of the cornea fitting into the sclerotica; 14, circular sinus of the iris; 15, choroidea; 16, the anterior portion, constituting the ciliary body; 17, the ciliary muscle; 18, the ciliary processes; 19, retina; 20, its origin; 21, the ora; 22, central retinal artery; 23, vitreous humor; 24, 25, 26, hyaloid tunic; 27, suspensory ligament of the crystalline lens; 28, 29, iris; 30, pupil; 31, posterior chamber, and 32, anterior chamber occupied by the aqueous humor.

FIGURE 10. (After Sappey.)

The iris separates this chamber into two parts, the *anterior chamber* and the *posterior chamber*. The two chambers communicate with each other through the pupil.

The aqueous humor is faintly alkaline, and has a specific gravity of about 1005.

The aqueous humor is rapidly reproduced after it has been evacuated, as occurs frequently in accidents or operations upon the eye.

The aqueous humor is secreted by the ciliary processes, and the posterior surface of the iris.

The aqueous chamber has its exit at the angle between the iris and cornea, through the spaces of Fontana (the meshwork of the ligamentum pectinatum) into the canal of Schlemm, thence through a system of valves into a plexus of veins where it is conveyed to the choroidal veins.

The aqueous chamber together with the canal of Petit and the canal of Schlemm can properly be regarded as lymph spaces, and the aqueous humor as the lymph occupying these spaces.

The *crystalline lens* is a lens-shaped body occupying the space in the antero-posterior diameter of the eye, between the aqueous humor and the crystalline humor. It is perfectly transparent and very elastic.

The crystalline lens is a nonvascular agglomeration of transparent fibrils, arranged in radiating sectors, and enclosed in a fibrous capsule (the lens capsule), situated in the hyaloid fossa of the vitreous humor, and held in position by the annular suspensory ligament, (zonule of Zinn).

The crystalline lens after the age of twenty-five years becomes hardened in the center, which portion is termed the nucleus, which is surrounded by the non-sclerosed portion or cortex.

The crystalline lens, owing to its shape, is the most refracting of the media of the eye, its action being analogous to that of the convex lenses of optical instruments. It causes the rays of light passing through it to focus, in the normal eye, upon the retina.

Owing to the action of the ciliary muscles the crystalline lens varies in its convexities, owing to the distance at which the gaze is directed. When the object viewed is near, the lens is more convex; when it is at a distance, it is less convex.

The crystalline lens in the adult is about 8.5 mm. in its transverse diameter and about 6.5 mm. in its antero-posterior diameter.

At about forty-five years of age the crystalline lens loses so much of its elasticity and becomes so hard that a convex lens must be placed before the eye to enable it to discern near objects plainly.

The absence of the crystalline lens is known as *aphakia*. This condition has never been discovered as a congenital defect, but as the result of an accident or surgical operation.

The *vitreous humor* is the transparent gelatinous mass which occupies the posterior cavity of the eye.

The vitreous humor is not a secreted fluid like the aqueous, but an embryonic product which is formed in the vitreous chamber at a very early period of fœtal life. When a portion of it is lost, it therefore is not reproduced, but its place is filled by lymph, which does not cause appreciable loss of vision.

The vitreous humor is surrounded by a transparent capsule, the *hyaloid membrane*.

The anterior portion of the vitreous humor has a deep depression, the *hyaloid fossa*, in which the posterior surface of the crystalline lens rests.

The vitreous humor is traversed in its anteroposterior diameter by a canal, the *canalis hyaloideus*, which has its origin at the papilla of the optic nerve, and extends to the posterior surface of the lens, and is occupied during fœtal life by the hyaloid artery.

The vitreous humor in the fully developed eye is destitute of vessels, the hyaloid artery which occupied it during fœtal life having formed the retinal vessels, and is dependent upon the surrounding tissues, namely the tunica vasculosa, for nutrition.

The principal function of the vitreous humor is in connection with the aqueous, to hold the choroid, the retina and the sclerotic in their respective situations. Were it not for these humors, and especially the vitreous, the eye would collapse.

THE BLOOD VESSELS OF THE EYE AND ITS APPENDAGES.

The vascular supply of the eyeball and its appendages is affected mainly from branches of the ophthalmic artery as follows:

Lachrymal,
Supra-orbital,
Superior-palpebral,
Inferior-palpebral,
Nasal,
Short-ciliary,
Long-ciliary,
Superior-muscular,
Inferior-muscular,
Inferior-orbital, (branch of the internal maxillary),

Anterior-cerebral, (branch of the internal carotid).

The *lachrymal artery* supplies the lachrymal glands, the conjunctiva, and, inosculating with the superior palpebral, assists in supplying the upper lids.

The *supra-orbital artery* supplies the superior rectus and levator palpebra.

The *superior palpebral artery* supplies the upper lid. It entirely encircles it, forming an arch at the free margin where it lies between the orbicularis palpebra muscle and the tarsal cartilage.

The *inferior palpebral artery* supplies the lower lid, and encircles it in the same manner as the superior does the upper lid.

The *nasal artery* supplies the lachrymal sac.

The *short ciliary arteries* enter the sclerotic around the optic nerve entrance, and supply the choroid and ciliary processes. (They number from twelve to fifteen.)

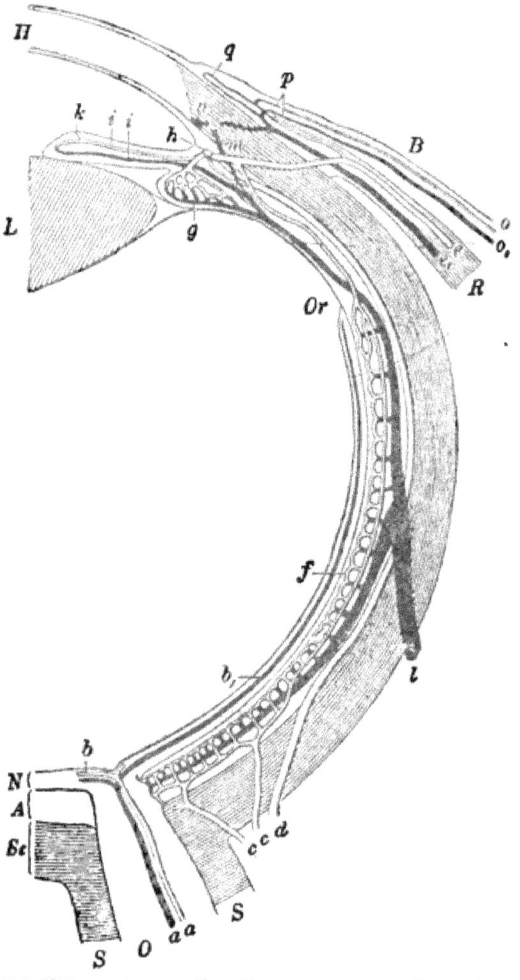

BLOOD-VESSELS OF THE EYE (SCHEMATIC, AFTER LEBER).

The *retinal system of vessels* is derived from the central artery, a, and the central vein a_1, of the optic nerve, which give off the retinal arteries, b, and the retinal veins b_1, These end at the ora serrata. *Or*

The *system of ciliary vessels* is fed by the posterior short ciliary arteries, c, c, the posterior long ciliary arteries, d, and the anterior ciliary arteries, e. From these arise the vascular network of the chorioidal capillaries, f, and of the ciliary body, g, and the circulus arteriosus iridis major h. From this last spring the arteries of the iris, i, which at the smaller (inner) circumference of the latter form the circulus arteriosus iridis minor, k. The veins of the iris i_1, of the ciliary body, and of the chorioid are collected into the vasa vorticosa, l; those veins, however, that come from the ciliary muscle (*m*) leave the eye as anterior ciliary veins, e_1. With the latter, Schlemm's canal, n, forms anastomoses.

The *system of conjunctival vessels* consists of the posterior conjunctival vessels, e and o_1. These communicate with those branches of the anterior ciliary vessels which run to meet them—that is, with the anterior conjunctival vessels, p—and form with these the marginal loops of the cornea g. *O*, optic nerve; *S*, its sheath; *Sc*, sclera; *A*, chorioid; *N*, retina; *L*, lens; *H*, cornea; *R*, internal rectus; *B*, conjunctiva.

FIGURE 11.

The *long ciliary arteries* (two in number) enter the sclerotic on its posterior surface on each side of the optic nerve, and pass forward on each side of the eyeball between the sclerotic and choroid as far as the ciliary ligament, where they divide into two branches and form the *circulus iridis major* at the ciliary margin, and the *circulus iridis minor* at the pupillary margin of the iris.

The *superior muscular artery* supplies the blood for the levator palpebra, superior rectus and superior oblique muscles.

The *inferior muscular artery* is distributed to the external rectus, inferior rectus, and inferior oblique muscles.

The *infra orbital artery* supplies the inferior rectus and inferior oblique muscles, and the lachrymal gland. (This latter branch is often wanting.)

SECTION VIII.

THE LYMPHATIC SYSTEM.

The lymphatic system of the eye consists of the *lymph channels* and *lymph spaces*.

The lymphatic system of the eye is divided into two portions, viz,—the *anterior lymphatic system* and the *posterior lymphatic system*.

The anterior portion occupies two large spaces, viz,—the *anterior* and *posterior chambers* of the eye.

These two chambers are simply lymph spaces and communicate by means of the pupil.

The lymph in the anterior portion of the eye is secreted by the iris and ciliary processes. The greater portion of the lymph exudes from the posterior surface of the iris and the anterior surface of the ciliary body. A small amount comes from the corneal endothelium.

The lymph from the anterior portion of the eye is discharged from the anterior chamber by being filtered through the sieve work of the *ligamentum pectinatum* into the *canal of Schlemm*. From the canal of Schlemm the lymph is carried into the ciliary veins.

The posterior portion of the lymphatic system of the eye occupies the *hyaloid canal*, the *peri-choroidal space* and *Tenon's space.*

The lymph supplying the hyaloid canal is derived from the blood vessels of the optic nerve, that supplying the peri-choroidal space from the blood vessels of the choroid, and that supplying Tenon's space from the blood vessels supplying the optic nerve and Tenon's capsule.

The outflow of lymph from all of the above mentioned spaces is through the lymph passages which spread out along the optic nerve.

The lymph passages through which the outflow takes place are the *intravaginal space* and the *supravaginal space.*

As most of the lymph leaves the eye through the anterior lymph passages, the anterior system is the most important factor in many affections of the eye.

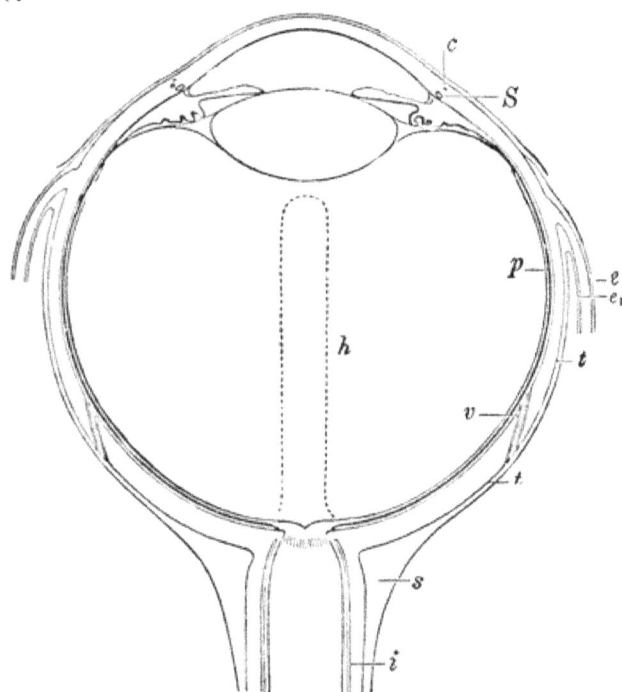

LYMPH-PASSAGES OF THE EYE. SCHEMATIC.

S, Schlemm's canal ; *c*, anterior ciliary veins ; *h*, hyaloid canal ; *p*, perichorioidal space, which communicates by means of the venae vorticosae, *v*, with Tenon's space ; *t, t* ; *s*, supravaginal space ; *i*, intervaginal space ; *e*, *e₁*, continuation of Tenon's capsule upon the tendons of the ocular muscles lateral invagination .

FIGURE 12. After Fuchs.

A thorough study of the lymphatic system of the eye should be made, for upon all theories of intra-ocular pressure it has important value.

THE MUSCLES OF THE EYE.

The eyeball is moved in various directions by the actions of its muscles.

The muscles which contribute to its movements are the *external rectus*, the *internal rectus*, the *superior rectus*, the *inferior rectus*, the *superior oblique* and the *inferior oblique*. These muscles are called the *extrinsic* muscles of the eye.

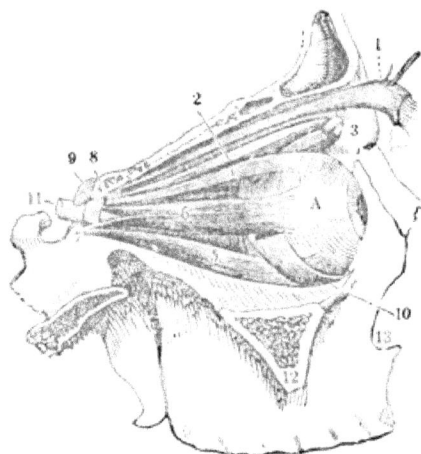

MUSCLES OF THE EYE. 1, the palpebral elevator; 2, the trochlear muscle; 3, the pulley through which the tendon of insertion plays; 4, superior rectus muscle; 5, inferior rectus muscle; 6, external rectus muscle; 7, 8, its two points of origin; 9, interval through which pass the oculo-motor and abducent nerves; 10, inferior oblique muscle; 11, optic nerve; 12, cut surface of the malar process of the superior maxillary bone; 13, the nasal notch. A, the eyeball.

FIGURE 15. (After Morton.

The *external rectus muscle* arises from two heads, the upper one from the outer margin of the optic foramen, and its lower head from the lower margin of the sphenoidal fissure, and is inserted into the sclerotic coat about 6 mm. from the margin of the cornea. It is the longest of the recti muscles.

The *internal rectus* arises with the inferior rectus from a common tendon from the lower and inner circumference of the optic foramen, and is inserted into the sclerotic about 6 mm. from the margin of the cornea. This muscle is the broadest and strongest of the recti.

The *superior rectus muscle* arises from the upper margin of the optic foramen, and from the fibrous sheath of the optic nerve, and is inserted into the sclerotic about 6 mm. from the margin of the cornea. This muscle is the thinnest and narrowest of the recti.

The *inferior rectus muscle* arises by a common tendon with the internal rectus, and is inserted into the sclerotic at about 6mm. from its corneal margin, as the other recti muscles.

The *superior oblique* arises from the inner margin of the optical foramen and terminates in a rounded tendon which glides through a cartilaginous pulley beneath the internal angular process of the frontal bone, then it passes under the superior rectus and is inserted into the sclerotic midway between the margin of the cornea and the optic nerve, between the superior and the external recti muscles.

The *inferior oblique* arises from the orbital plate of the superior maxillary and is inserted into the sclerotic between the superior and external recti.

The varied movements of the eye are accomplished by these muscles, for the performance of which they are arranged in three pairs, each pair consisting of two muscles, the actions of which are antagonistic.

The *first pair* is the rectus internus and rectus externus, the former rotating the eye inward and the latter outward.

The *second pair* is the rectus superior and the rectus inferior, the rectus superior rotating the eye upward and the rectus inferior downwards.

The *third pair* is the superior oblique and the inferior oblique, the former inclines the vertical motion of the eye inward (wheel motion inward) and the latter inclines the vertical meridian outward (wheel motion outward).

NERVE SUPPLY OF THE ORBITAL MUSCLES.

The external rectus is supplied by the sixth cranial or abducens.

The internal rectus muscle is supplied by the third cranial or motor oculi.

The superior rectus is supplied by the third cranial or motor oculi.

The inferior rectus is supplied by the third cranial or motor oculi.

The superior oblique is supplied by the fourth cranial or patheticus.

The inferior oblique is supplied by the third cranial or motor oculi.

THE BLOOD VESSELS OF THE OCULAR MUSCLES.

The muscles are all supplied by branches from the ophthalmic artery.

The veins of the muscles empty into the ophthalmic and facial.

CHAPTER II.

DISEASES OF THE ORBIT.

THE following may be enumerated as the diseases of the orbit:

Periostitis, necrosis, cellulitis, tumors.

Periostitis of the orbit may be *acute* or *chronic.*
Acute periostitis may be *local* or *diffused.*

Acute periostitis is characterized by pain, swelling of the lids, inflammation of the conjunctiva, slight protrusion of the ball, and tenderness over the seat of the affection if it is localized.

The rim of the orbit is attacked with special frequency on account of its exposure to traumatism.

If the inflammation is diffused, the pain, swelling of the lids and conjunctival inflammation is much more pronounced. In the severe cases there is an increase of temperature, and occasionally delirium.

As acute periostitis of the orbit has many symptoms in common with orbital cellulitis, a differential diagnosis is often difficult, especially as the latter generally accompanies the former.

The absence of tenderness on pressure upon the margin of the orbit is considered by many oculists as indicative of absence of periostitis.

There is not so much tenderness and swelling

of the lids or conjunctival irritation in periostitis
as in cellulitis.

In acute periostitis there is always a tendency
to suppuration, but there is no tendency to suppuration in chronic periostitis.

Chronic periostitis of the orbit is accompanied
by pain in and about the orbit with tenderness on
pressure of the ball, but swelling of the lids, chemosis of the conjunctiva and the exophthalmus
which are present in the acute variety are usually
absent in the chronic.

Periostitis of the orbit is usually due to some
traumatism, as blows or penetrating wounds near
the margin.

The chronic variety is usually the result of
syphilis or rheumatism.

Treatment, in the acute variety, consists in the
application of fomentations, as hot as can possibly be borne, in order to hasten resolution and
absorption.   If impossible, then evacuate the pus
at the earliest possible moment.

In the chronic variety we must depend on constitutional remedies.

NECROSIS OF THE BONES OF THE ORBIT.

As *necrosis* of the bones of the orbit usually
follows periostitis, the affection rarely occurs primarily.   We have therefore in this disease all of
the accompanying symptoms, as pain and swelling
of the lids and conjunctiva.

As in periostitis, the rim of the orbit is usually the seat of the affection.

Accompanying the pain is a circumscribed swelling, indurated at first, followed usually by softening, then perforation and evacuation. Absorption and resolution rarely, though occasionally, take place. After evacuation the sound, if introduced into the cavity, will come in contact with rough bone.

When the necrosis is situated deep in the orbit, there is increased inflammatory action. The lids become very swollen, the conjunctiva chemotic, and the ball protruded to an exceedingly disfiguring extent. The pain is very severe and usually worse at night.

The cellular tissue cannot remain intact in periostitis and necrosis, but is more or less implicated and often becomes so seriously involved that the eye is sacrificed.

The causes of necrosis, like that of its predecessor, periostitis, are usually wounds and injuries of the orbit (mostly at its rim), rheumatism and syphilis. Occurring in children, it is usually of scrofulous origin.

If the rim of the orbit is alone affected, the prognosis is usually good.

If however the necrosis is deep in the orbit, the prognosis is grave, because of the extension of the inflammation to the meninges of the brain, the injury to the optic nerve, and the involvement of the eye in the general inflammatory condition.

In the *treatment* at the incipiency of the affection, no matter what the cause, warm fomentations are indicated.

The applications of tincture of iodine, mercurial ointment and other medicines said to have deobstruent properties are useless.

If resolution and absorption cannot be effected, the escape of pus must be hastened by incision with a narrow scalpel, piercing the conjunctiva as near the fornix as possible, and following the orbital wall to the abcess. The wound should be thoroughly cleansed with antiseptic solutions—preferably the bichloride of mercury solution, 1-5000th.

If the patient is rheumatic or has had syphilis, constitutional treatment is necessary. My preference is the iodide of potassium. In the adult from 30 to 60 drops of the saturated solution, in three or four tablespoonfuls of milk, before meals.

Orbital Cellulitis.

The cellular and fatty tissue forming the cushion for the eyeball is liable to inflammation.

The symptoms accompanying orbital cellulitis are fever, pain, œdema of the lids, inflammation of the conjunctiva, frequently chemosis and more or less protrusion of the eyeball.

The protrusion of the eye is frequently so great that the lids will not cover the entire ball, and the movements of the globe are limited, often to a considerable extent.

If pressure is made between the margin of the orbit and the ball, the tumor may be located, which is generally at the outer and upper angle of the orbit. The tumor gradually increases in size and fluctuation is finally detected; then follows suppuration by an opening either in the integument or the conjunctiva.

As a rule, in cellulitis there is a tendency to complete recovery; however, in some cases the inflammatory action is so great that the ball partakes of the general inflammation, and as a result there is occasionally a loss of vision from atrophy of the optic nerve, suppuration of the cornea, or disorganization of the deeper structures of the eye.

Occasionally cellulitis is fatal, death resulting from pyæmia, meningitis or cerebral abscess.

The causes of orbital cellulitis are colds, traumatism, extension of inflammation from contiguous parts (as in erysipelas), tuberculosis and syphilis (generally from syphilitic caries of the bones).

Treatment.—First warm fomentations, consisting of water alone, as hot as can be borne, and changed often. Never apply poultices of any kind about the eye. Then as early an evacuation of the pus as possible.

TUMORS OF THE ORBIT.

Tumors of the orbit consist of *cystic tumors*, **vascular** *tumors* and *malignant tumors*.

1. *Cystic Tumors.*

The most frequent of these is the *dermoid cyst*, a congenital tumor, but which often grows to considerable size after birth.

The *dermoid cyst*, although classified as an orbital tumor, cannot be said strictly to be such, as it is always found in the integument, not within the orbit, but at its upper and outer, or upper and inner angle.

The dermoid cyst consists of subaceous matter.

In removing the cyst care must be taken to dissect out the entire cyst, for even a part of it remaining behind is liable to give rise to a recurrence of the tumor.

2.   *Vascular Tumor.*

The different forms of vascular tumor, rarely, although occasionally, occur in the orbit.   They are *aneurism* and *angiomata* of the orbit.

The *aneurism* consists of dilated vessels, and the *angiomata* of new growths consisting principally of blood vessels.

The angiomata consists of two forms, the *teleangiectasis* and the *tumores cavernosi.*

The *teleangiectasis* are congenital, and consist of bright red spots in the integument of the lids, from which it gradually extends to the orbit.

The *tumores cavernosi* lie beneath the integument of the lids, which is protruded forward, through which is seen the bluish lustre.

Cavernous tumors usually develop in the orbit

first, and growing, gradually push the eye-ball forward.

Vascular tumors of the orbit vary in size. Active bodily exercise, and the acts of crying, coughing, etc., increase their volume, while pressure upon the eye toward the orbit diminishes their size.

If the size of the tumor increases to such an extent as to endanger the eye, surgical procedure must be instituted.

If the tumor is sharply defined and is encapsulated, it should be removed.

If diffused, electrolysis is indicated.

3. *Malignant Tumors of the Orbit.*

The *malignant tumors* of the orbit may be classed as the *sarcomata* and *carcinomata*.

The *orbital sarcomata* take their origin in the orbit proper, that is from the bone and periosteum, from the cellular tissue, the optic nerve and its sheaths, and from the lachrymal gland.

Orbital sarcomata are enclosed in a covering of connective tissue. They are usually rounded and soft in consistence.

There is a form of sarcomata that develops primarily in the eyeball. They usually have their origin in the choroid or retina, and after bursting through the sclerotic, fill the orbit and push the ball forward.

The *carcinomata* of the orbit do not usually have their origin in that cavity, but in the conjunctiva or the lids, and extend backward.

Carcinomata of the orbit proper have their origin in the lachrymal gland.

All malignant tumors are of serious import, no matter where the origin of the affection, whether in the orbit proper, in the lids or conjunctiva, or within the eyeball itself.

If operative treatment is successful (and no other is), it should be resorted to at a very early date.

EXOPHTHALMIC GOITRE.

Exophthalmic goitre (Basedow's Disease; Grave's Disease) is a condition in which there is more or less prominence of the eyes and with which there is enlargement of the thyroid gland, and paroxysmal palpitation of the heart.

Exopthalmic goitre is not a disease *per se* of the eyes or the orbit, but is due to a condition of the general system in which the brain, heart, and the alimentary tract are more or less implicated.

The student is referred to treatises on general medicine for the consideration of this subject.

CHAPTER III.

SECTION I.

THE DISEASES OF THE LACHRYMAL GLAND.

THE lachrymal gland is subject to the following affections: *Inflammations, Hypertrophy, Cancerous Growths,* and *Fistulae.*

Inflammation of the lachrymal gland may be *acute* or *chronic.*

Acute inflammation of the lachrymal gland is indicated by severe darting pain extending from the orbit over the forehead and side of the head. The conjunctiva becomes much congested and chemotic, and the lids are very oedematous. The globe of the eye is pressed downwards and forwards, or inwards and backwards.

Like all inflammatory actions, it terminates either in resolution or suppuration. If, in resolution, the gland is gradually reduced in size, and there is general subsidence of the swelling and congestion of the adjacent tissues.

If the inflammatory condition ends in suppuration, then there will be felt fluctuation at the upper and outer part of the orbit, which after a time will burst and discharge through one or more openings in the upper lid.

In the *chronic* form, the periosteum and the bone in the vicinity of the gland become involved, which usually results in a fistula, which remains open as long as there remains diseased portion of bone.

Treatment in the early stages of the acute form consists in the application of cold compresses to the parts affected.

If suppuration is inevitable, then the evacuation of the pus at the first evidences of fluctuation. After the contents are thoroughly discharged, the wounds must be antiseptically dressed.

The more thoroughly the pus sac is cleaned, the less liable are we to have periostitis and necrosis of the bones of the orbit. It is mostly the neglected cases of suppuration of the lachrymal gland that are followed by periostitis and necrosis.

Hypertrophy of the lachrymal gland is indicated by the protuberance of the parts in the vicinity of the gland and may be easily felt behind the outer part of the upper lid.

It is painless, nodular and increases in size very slowly and, like acute inflammation, may gradually disappear or may suppurate and result in a chronic periostitis or necrosis of the bones of the orbit.

In the treatment of hypertrophy of the lachrymal gland, absorption is often effected by the local application of the tincture of iodine. That this treatment may be most effective, the integument over the gland must be kept irritated from its use. At the same time constitutional remedies, as the iodide of iron, fresh air, and good nourishing food must be relied upon in the hope of promoting the absorption of the tumor.

In case of chronic enlargement of the gland, it may be removed. Should the gland suppurate, it should be treated on the same aseptic principles as mentioned in acute suppurative inflammation.

The time is past for hastening suppuration by hot poultices and the like. At the first indication of suppuration the parts must be laid open, and the contents evacuated, and the pus sac thoroughly cleansed with antiseptic solutions.

Cancerous growths of the lachrymal gland are occasionally met, and are *sarcomatous* or *carcinomatous*.

In *sarcoma* of the lachrymal gland the growth is painless and usually slow.

In *carcinoma* the growth is also slow and painless, but other glands of the body become enlarged and evince the malignancy of the affection.

The *treatment* for cancerous growths is removal as soon as the diagnosis is established.

The gland can be removed without serious detriment to the eye.

Fistula of the lachrymal gland is the result of an abscess or an injury. It is easily diagnosed by the clear fluid which constantly drains from it.

The treatment is to establish drainage into the conjunctival sac, the best manner of which is to pass a probe along the course of the fistula, then evert the lid while the probe remains, and cut down upon the probe through the conjunctival surface. In this way another fistulous opening will be established and the secretion thus conducted to its proper destination.

The mouth of the fistulous opening on the skin should be cauterized, and the wound dressed antiseptically.

SECTION II.

DISEASES OF THE LACHRYMAL APPARATUS.

The *lachrymal apparatus*, consisting of the *canaliculi*, the *sac*, and the *ducts*, is liable to the inflammations usual to mucous surfaces, the chief effect of which is an impediment to the transmission of the secretions from the eye into the nasal cavity. When such an obstruction exists, the condition is known as *epiphora*, *stillicidium lachrymarum*, or *watery eye*.

The causes of obstruction of the lachrymal passages are numerous; the puncta may be displaced as from an ectropion, so that the tears cannot enter, or the canaliculi may be obstructed from an inflammation of the mucous membrane of that part of the passage, or the obstruction may occur from like causes in the sac or ducts.

The lachrymal sac is occasionally subject to acute inflammation, the result of which is an abcess.

This condition is known technically as *dacryocystitis*.

Acute inflammation of the lachrymal sac is usually the result of a conjunctivitis or a catarrh of the nasal passages.

This is an acute suppurative inflammation of the submucous connective tissue, conspicuous by the sudden development of a highly sensitive phlegmonous swelling. This swelling, more marked in the region of the lachrymal sac, is more or less diffused, and the inflammation frequently assumes an erysipelatous type. The swelling for the first forty-eight hours is very painful, and the tumefaction becomes more and more tense, until the height of the inflammatory action has been reached.

The pain then somewhat subsides, the integument at the apex of the swelling becomes somewhat yellowish in appearance, and finally gives way at this point, and the contents, a large quantity of pus, are evacuated.

Upon the evacuation of the pus, the swelling rapidly recedes, the pain ceases, and all evidences of inflammatory action gradually subside.

It will be fortunate if this is the termination of the difficulty, for the effects are often of serious consequence, especially if the wound should not heal, and a fistula should be formed, in which case the tears, instead of passing through the sac into the ducts, pass from the sac through this false passage and are discharged on the face

Stricture of the lachrymal duct, whatever the cause of it may be, usually results in an abcess of the lachrymal sac, and should the fistula close, the patient is liable to another attack of the inflammation; however, as long as the fistula remains open, there is no danger of its recurrence.

An *abcess of the sac* always injures it more or less, and should it recur very often, the mucous membrane lining the sac and the duct may be destroyed, or injured to such an extent as to permanently close the passage.

It is observed by some of the older writers that caries or necrosis of the lachrymal bone follows as a result of an abscess of the lachrymal sac, but a more recent and reasonable view is that the abscess is due to the necrosis of the lachrymal bone, which in turn is probably due to syphilis or tuberculosis.

Treatment.—Hot water fomentations as warm as can be borne should be employed, and as soon as the abscess is pointed, it should be freely opened. The old practice of opening the abscess by introducing a knife through the canaliculus into the sac is barbarous, and should not be resorted to, being painful in the extreme and productive of no special benefit.

After the inflammation subsides, an attempt should be made to open the stricture by means of a lachrymal probe, for in this lies our hope of preventing a recurrence of the abscesses, and if a lachrymal fistula exists, in causing the secretion to take its natural passage through the ducts.

OPERATION FOR SLITTING UP THE LOWER CANALICULUS.

A thorough acquaintance of the anatomy of the parts is requisite before attempting to pass a probe through the sac into the nasal duct

The parts should be thoroughly cocainized by dropping a four per cent solution into the eye every two minutes for the period of a quarter of an hour.

The patient seated with his head thrown back, and the operator behind tensely draws the lower lid outwards and downwards with one hand, and with the other introduces the point of a canaliculus knife vertically into the punctum. When the point has well entered the punctum, the knife is then turned and passed horizontally along the canaliculus until its point reaches the inner bony wall of the sac. During its passage the edge of the knife is to be turned toward the conjunctiva, so as to make the incision near the inner edge of the lid wthout cutting into its border.

FIGURE 14. 'Noyes-Stilling's Canaliculus Knife.'

When the point of the knife reaches the inner wall of the sac, the eyelid being still kept on the stretch, the knife is to be elevated from the horizontal to the perpendicular, and the canaliculus is thus divided well up to the sac

To Introduce the Nasal Probe.—The canaliculus having been slit, a probe should be passed through the stricture at once. The manner in which this is done is as follows: A small-sized Bowman probe, usually a No. 3, is first introduced, by stretching the lid and by passing its point horizontally along the opened canaliculus until it reaches the inner bony wall of the sac; the instrument is then turned vertically and gently pressed through the sac, when the point is directed outwards and backwards, it passes into the nasal duct, through which it enters the nasal cavity.

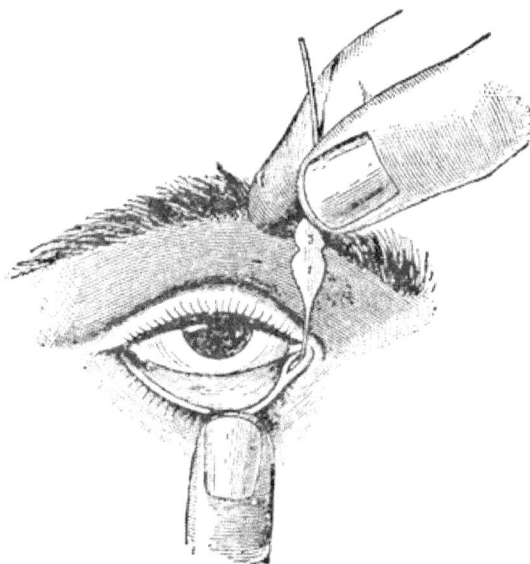

FIGURE 15. (After Meyer.)

If we find an obstruction that will not yield to a reasonably firm pressure, the probe must be removed and a smaller one, as a No. 2, and then if necessary a No. 1 introduced. It is seldom the case that the stricture is so firm that the smallest probe cannot be admitted, but it sometimes occurs that such is the case, in which event the stricture must be incised by means of a knife, which is introduced into the sac in the same manner as the probe and forced downward in the direction of the tube, through which it passes by firm pressure, if there are no bony obstructions.

After the lachrymal duct has been opened with the knife in the manner described, the probes must be passed at least once every forty-eight hours, the size being increased from time to time as the nature of the case demands.

In case there is a bony obstruction the situation is usually considerably complicated owing to the character of the obstacle, whether it is caused from an exostosis, or from the closure of the bony walls from an injury. If from the former all proceedings to open the duct may as well be discontinued, as the mucous membrane of the canal has probably become involved in the inflammation to such an extent that its usefulness has been permanently impaired. In case the walls are simply pressed together, as from an injury, a large dilator, as Webber's graduated dilator, can be pressed through, and the parts so separated that the tears will readily pass. One operation is usually sufficient, as in most cases the bones will remain if they are pushed back to their former position. If the walls should again come together, they should be pressed apart as before, until they permanently remain in such a position that the tears will freely pass through the duct.

It is sometimes necessary in addition to opening up the duct, to treat the sac by means of local medications. This is easily effected by means of the lachrymal syringe.

Any good astringent or antiseptic that is indicated can be injected into the sac by introducing the nozzle in the same manner as the probe, and when in place the fluid can be forced into the sac by pressure upon the handle of the piston.

Where there is a tendency to closure after the probe has been used a considerable time, a small silver or lead style of the proper length can be introduced and the upper portion bent so it can lie in the slit of the canaliculus unobserved, which should be allowed to remain a reasonable length of time, unless the parts should become inflamed, when it must be removed, and if necessary reinserted after the inflammatory action has subsided.

Mucocele is a tumor formed by the detention of mucous in the lachrymal sac, owing to an obstruction in the duct, from whatever cause. The tumor is non-inflammatory, varies in size, and is sometimes very large.

If pressure is made upon the tumor, the contents can be pressed out through the puncta, and in some cases through the duct into the nose.

The treatment of mucocele is the same as for any obstruction of the lachrymal passage, and consists in slitting up the canaliculus, and dilating the ducts with the probes.

CHAPTER IV.

SECTION I.

INJURIES AND DISEASES OF THE LIDS.

THE integument of the eyelids is liable to the injuries and diseases incident to the integument of the other parts of the body, as contusions, wounds, burns and scalds, abscesses, tumors, naevi, herpes, erysipelas, epithelioma, papillomita (warts), indurated chancre, etc., the treatment of which will be conducted on general principles.

Contusions and wounds of the eyelids are very common and are frequently accompanied by other more serious lesions of the globe or orbit.

Contusions of the lids may vary from a slight redness to a severe ecchymosis.

The absorption of the ecchymosis is hastened by the application of hot water compresses, and in most cases it is the only remedial means necessary.

All persons who suffer the inconvenience of a "black eye" seem very anxious to get rid of it as soon as possible; but there are no medicinal remedies that will hasten the absorption better than the hot water except the removal of the blood by the natural or the artificial leech; and this is always more or less disappointing to the patient, for the result is not as effective as wished for.

Incised and lacerated wounds should have their edges brought nicely and perfectly together, preferably by a fine suture. Do not trust to adhesive strips for the reason that they are liable to slip.

Strict attention must be given these cases in order to avoid a displacement of the puncta lachrymalis, or an entropion or ectropion.

While burns and scalds of the lids require the same general treatment that is given to like injuries on other parts of the body, special attenton must be given these cases to prevent inversion or eversion of the lids, their adhesion to each other or the adhesion of either lid to the ball. To prevent the lids from uniting together or the lid uniting to the ball, the eye should be examined every day, and any adhesions should be separated with a smooth probe or blunt hook.

SECTION II.

The diseases of the lids may be classified as *blepharitis, hordeolum, chalazion, dermoid cyst, xanthelasma, ptosis, ectropion, entropion, trichiasis* and *distichiasis, ankyloblepharon, symblepharon* and *epicanthus.*

Blepharitis—known as ophthalmia tarsi, tinea tarsi, sycosis tarsi—is an inflammatory affection of the border of the eyelids, induced by a chronic diffuse inflammation of the meibomian glands, sebaceous glands, and the follicles of the eye-lashes.

Blepharitis is divided into two distinct varieties, viz. *blepharitis squamosa* and *blepharitis ulcerosa.*

Blepharitis squamosa is characterized by an inflammatory condition of the borders of the lids, confined principally to the roots of the cilia and their vicinity, the affected parts of which are covered with scales like dandruff.

Upon the removal of these scales no ulcers are found, but simply a congested condition of the skin.

The scales are formed from the hyper-secretion of the sebaceous glands, which solidifies upon coming in contact with the air.

This affection may be confined to one lid or to all, but generally both lids of both eyes in time become involved.

Blepharitis ulcerosa is characterized by a greater degree of inflammatory action than in the former variety. The border of the lids is covered with a yellow crust, underneath which, after washing off, we find excavated ulcerations. These excavations are caused from abscesses which have originated from suppuration of a hair follicle and the sebaceous gland belonging to it.

Like in blepharitis squamosa, this variety may affect one lid or all.

The cilia become deficient and straggling, because their follicles and the sebaceous glands connected to them have been destroyed or otherwise seriously injured by the suppurative process.

Blepharitis ulcerosa is distinguished from the blepharitis squamosa by the yellow crusts which cover the border of the lid and glue the cilia together, and by its deep abscesses and the purulent form of the inflammation.

The yellow crusts are formed from the inspissated pus, and from the abscesses of the hair follicles and sebaceous glands.

Blepharitis squamosa is less inflammatory in its action than the ulcerosa and causes very little inconvenience to the patient more than the disfiguring effect due to the reddened border of the lids.

The following sequelae are the results of an unchecked blepharitis ulcerosa:

The eye-lashes become stunted and misplaced, and turn in and irritate the conjunctiva. This condition known as *trichiasis*.

The lashes may drop out, which condition is known as *madarosis*.

The lids may become slightly everted, showing the red, thickened edges, designated *lippitudo*.

The tears may overflow on account of displacement of the tear duct, *epiphora*.

Blepharitis varies from a simple congestion to the most severe inflammation of the edges of the lids, which become red and thickened and burrowed with deep ulcerations.

In severe cases of blepharitis ulcerosa, the ulcers become so deep, and the inflammation is so extensive, that not only the sebaceous glands and the hair follicles become destroyed, but the meibomian glands suffer to a greater or less extent.

Causes.—Granular conjunctivitis, catarrhal conjunctivitis, irritating causes, such as exposure to dust and smoke and artificial light, closure of the lachrymal duct. It frequently follows an attack of measles or other exanthemata.

The squamosa variety is often due to an error of refraction, which, when the cause is removed, promptly recovers.

Treatment.— This varies with the variety, the severity and the cause.

The cause, if ascertained, must always be removed, if possible.

If closure of the lachrymal duct, it should be opened. If due to an error of refraction, the vision should be corrected. Some of the worst cases of the squamous variety, of many years' standing, which I have ever witnessed, recovered immediately upon the patient's having the refractive error corrected.

The patient's general health, too, must be taken into consideration. If there is a tendency to a strumous diathesis, the general treatment for such cases, as iron, cod liver oil, quinine, etc., are indicated.

Although most authors claim this to be a disease principally of childhood, it has not been my experience to find it so by any means. On the contrary, I find more cases of both varieties in the adult.

The local treatment is simple. Remove the incrustations with an alkaline solution (bicarb. soda 10 grs., aqua 1 oz.) applied by means of cotton, until well saturated, when the crusts can be easily rubbed off with a small piece of dry cotton. If there are ulcers, apply once daily a solution of nitrate of silver (10 grs. to 1 oz. of distilled water) with a small cotton applicator, being cautious not to allow any of the solution to enter the eye. Immediately after this apply the ointment of the yellow oxide of mercury, 1 grain to 1 dram vaselina.

It is sometimes necessary in blepharitis ulcerosa to remove the cilia. This should not be done unless there are abscesses at their roots. The removal in such cases gives exit to the pus, and gives us the opportunity to make application of the proper remedies to the parts affected.

Recovery, especially in the ulcerosa variety, is protracted.

HORDEOLUM.

Hordeolum (commonly known as stye). A suppurative inflammation of one of the sebaceous glands of the lids.

If the inflammation is connected with the glands of the cilia, the Zeissian glands, the affection is termed *hordeolum Zeissianum,* or external stye. In this case, the abscess points around one of the lashes.

If the abscess is seated in a meibomian gland, it is termed *hordeolum Meibomianum,* or internal stye, and it will open either to the border of the lid or the conjunctiva, but rarely to the skin.

The *treatment* consists in hot fomentations, in order to hasten the formation of pus, then free evacnation. When styes are recurrent, it is an indication that the general health is at fault and requires attention. The refractive condition of the eye should always be examined in recurrent styes. I have known some of the most stubborn cases to be relieved at once upon the correction of refractive error.

CHALAZION.

Chalazion, a tumor of the lid formed by the distension of one of the meibomian glands. Its envelope is formed by the walls of the gland and its contents are the product of its secretion. The tumor is immovable and varies in size from a pin's head to that of a pea.

Chalazion seldom leads to spontaneous perforation, although occasionally this happens. When spontaneous perforation occurs, it is usually on the conjunctival surface.

As the conjunctival surface is the point of least resistance, the tumor usually points on its surface, at which place there is a small dusky spot.

The cause of these tumors is a chronic inflammation of the meibomian gland or its excretory duct. If in the former, the tumor is caused from a hypersecretion of the gland. If from the latter, the duct is closed and there is a distention from an accumulation of the natural secretions.

The tumor is to be removed from the conjunctival surface of the lid by an incision into it. The contents, which is usually a soft gelatinous or sebaceous mass, should be thoroughly scooped out, and the cyst wall thoroughly broken up by a cystotome or other suitable instrument.

The cavity usually fills up with blood, which will in a few days absorb. Hot water fomentations will hasten the absorption.

The particular point in this operation is the thorough breaking up of the cyst, otherwise it will refill and the tumor again appear. The application of medicine to these tumors is entirely useless.

SECTION III.

DERMOID CYST.

A *dermoid cyst* is a congenital cyst, containing skin, hair or other structures of the cuticle.

Dermoid cyst is not a disease of the lids *per se*, other parts of the body being liable to it; but as it is oftenest found at the outer angle of the orbit, on or near the brow, it is here classified under the diseases of the lids.

Treatment.— An incision, parallel with the lid border, should be made through the integument, over the tumor, through which the cyst and its contents should be removed.

The wound should be thoroughly cleansed, the edges finely sutured, and antiseptically dressed.

XANTHELASMA.

Xanthelasma is a disease characterized by the development upon the integument of the eyelids of yellow or brown growths.

These growths are caused from a fatty degeneration of the skin, and vary in size and shape.

Treatment.—The treatment consists in **excision** or electrolysis, preferably excision.

PTOSIS.

Ptosis is characterized by a drooping of the upper lid, due to paralysis, atrophy, or absence of the levator palpebra superioris.

The levator palpebra superioris is supplied by a branch of the third nerve. Paralysis of this muscle alone is, however, exceedingly rare, as there is usually complete paralysis of the third nerve, especially when there is ptosis.

In case there is complete paralysis of the third nerve, the following muscles, being supplied by it, are affected: levator palpebra superioris, internal rectus, superior rectus, inferior rectus, inferior oblique, sphincter of iris, and the ciliary muscle.

In the above case then, beside having drooping of the lid, the eye would turn out, the pupil would be dilated, and the patient could not accommodate.

The levator palpebra superioris may become atrophied or enfeebled by a long continued inflammatory condition of the lid, notably trachoma.

There are also cases of congenital ptosis, wherein the muscle is defective, or entirely absent.

In old persons, ptosis may be caused from an excess of integument of the lids, together, often, with hypertrophy of the cellular tissue.

Ptosis, from paralysis, should be treated by electricity. Should this fail to establish innervation, surgical interference may become necessary. Excess of integument may be remedied by excision.

Treatment.—The integument is neatly excised along a line of the fold, usually in a horizontal direction, together with any hypertrophied tissue which may exist, after which the edges are neatly united by as many sutures as are necessary to obtain perfect coaptation of the edges of the wound.

All operations upon the lids, entailing a loss of integument, must be carefully made, so that the normal contour of the lids may not be interfered with to any great extent. The removal of too much integument, resulting in an ectropion, would be a serious consequence.

ECTROPION.

Ectropion is that condition of the eyelid in which it is everted, or turned out away from the eye ball, and the conjunctival surface exposed.

This condition generally affects the lower lids. It is mostly caused by contraction of the integument of the lids, as a result of wounds from burns and other injuries. It frequently follows blepharitis marginalis. Ectropion in aged persons is frequently due to atrophy of the orbicularis palpebrarum.

It affects mostly the lower lids, but the upper lid may also suffer from the same condition.

Ectropion may be *partial* or *complete*.

Ectropion is *partial* when only a part of the lid is everted.

Ectropion is *complete* when the entire edge of the lid is everted.

Ectropion is also *acute* or *chronic*.

Ectropion is *acute* when due to spasmodic contraction of the orbicularis palpebrarum in ophthalmia neonatorum, or other conditions, in which there is great tumefaction of the conjunctiva.

Chronic ectropion is usually caused by wounds of the integument of the lids. It also follows severe forms of blepharitis marginalis, and paralysis or atrophy of the orbicularis palpebrarum in senility.

The most annoying complication in ectropion is the irritation of the conjunctiva from its exposure to the air and dust, and the epiphora caused by the displacement of the puncta.

Treatment.—The treatment for ectropion varies with the condition. In the acute form, the inflammation must be reduced on general principles. If there is much chemosis, excision or scarification of the protruded conjunctiva gives great relief. Also, compresses of cotton, dipped in very warm water, as warm as can be borne, will prove valuable in reducing the inflammation.

The various operations recommended in ectropion are applicable only in the chronic variety. However, many of the worst looking cases of this variety will yield to medical treatment, especially the ectropion of the lower lid of old persons, where the conjunctiva is thickened and the tissues relaxed.

In cases of this kind the exposed conjunctiva should be touched, once a week, with the solid stick of argenti nitras.

The operator should have at hand a saturated solution of sodium chloride, to immediately apply to the parts to which had been applied the argenti nitras.

This application is for the purpose of neutralizing the effect of the nitrate of silver, so that its irritating influence will not be carried farther than intended. A few applications of this kind will generally correct the most intractable ectropions.

There are many operations recommended for ectropion, the most important of which are the Argyle Robertson method, the Adams operation, Wharton Jones operation, and many others, which are modifications of those mentioned.

The Argyle Robertson method is adapted, principally, to those long-standing cases due to thickened conjunctiva and relaxed tissue, wherein the tarsal cartilage loses its normal shape.

The following is his description as given in the *Ophthalmic Review*, February, 1884:

"The materials required are:

"1.   A piece of thin sheet-lead about 1 inch
long and ¼ inch broad, rounded at its extremities,
and with its cut margins smoothed.   This piece of
lead must be bent with the fingers to a curvature
corresponding to that of the eyeball.

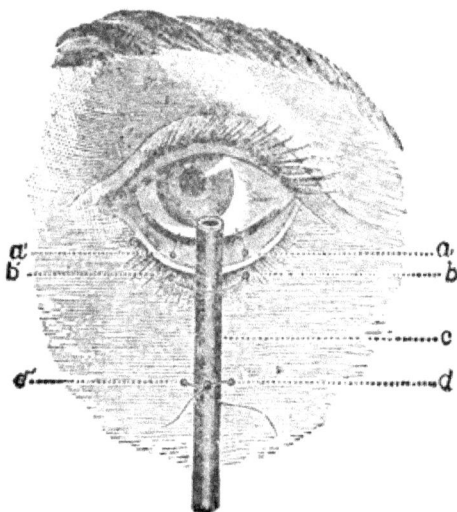

FIGURE 16.   After Swanzy.)

"2.   A waxed silk ligature about 15 inches long,
to either extremity of which a long moderately
curved needle is attached.

"3.   A piece of fine india-rubber tubing (the
thickness of a fine drainage-tube).

"The operation is performed by perforating the
whole thickness of the lid with one of the needles
at a point (b) one line from its ciliary margin, and
a quarter of an inch to the outer side of the center of
the lid.

"The needle having been drawn through (at *a*), is passed directly downward over the conjunctival surface of the lid, till it meets the fold of conjunctiva reflected from the lid on to the globe, through which the needle is thrust — the point being directed slightly forward—and pushed steadily downward under the skin of the cheek, until a point (*d*) is reached about 1 inch or 1¼ inch below the edge of the lid, when the needle is caused to emerge, and the ligature is drawn through. The other needle is, in like manner, thrust through the edge of the lid at a corresponding point (*b¹*) a quarter of an inch to the inner side of the middle of the lid, then passed over the conjunctival surface of the lid through the oculo-palpebral fold of conjunctiva, and downward under the skin, till the point emerges at a spot (*d¹*) a quarter of an inch outward from the point of emergence of the first needle (*d*).

"The ligature is kept slack, or is slackened so as to permit of the piece of lead being introduced under the loops of the ligature that pass over the conjunctival surface of the lid, and of the piece of india-rubber tubing (*c*) being slipped under the loop at the edge of the lid (between *b* and *b¹*). The free ends of the ligature are now drawn tight, and tied moderately tightly over a lower part of the india-rubber tube. The excess of india-rubber is cut off—about a quarter of an inch beyond the ligature—and the operation is complete."

Adam's Operation:

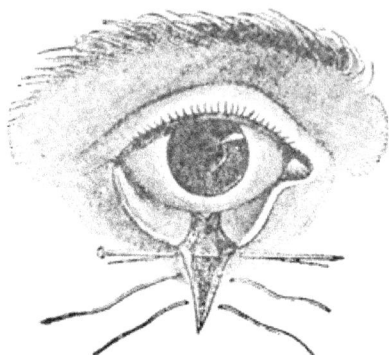

FIGURE 17. (After Meyer.)

Adam's method is adapted to such cases of ectropion as are due to cicatricial changes in the integument and tissues of the lid.

It consists in excising a triangular piece through the entire thickness of the lid, and bringing the lid into its normal position by fine sutures. Care must be taken not to excise too much, as the result may be contrary to that which was desired. (Adam's Operation).

Wharton Jones' Operation:

This operation is applicable principally to the lower lid, where the ectropion is caused by a cicatrix of the integument.

The cicatrix is included in two convergent incisions beginning near the angles of the eye, and uniting on the cheek in the shape of a V. The integument included in the incision is dissected up and made thoroughly movable. After this, the lid is restored to its normal position and the lips of the wound finely sutured, when the edges will present a Y shape. (Fig. 18 and 19).

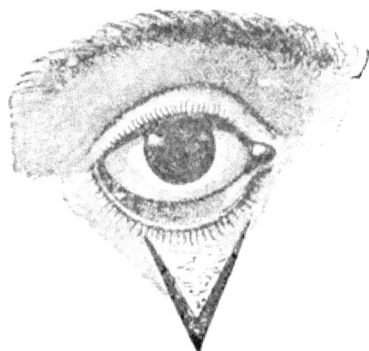

FIGURE 18. (After Meyer.) FIGURE 19. (After Meyer.)

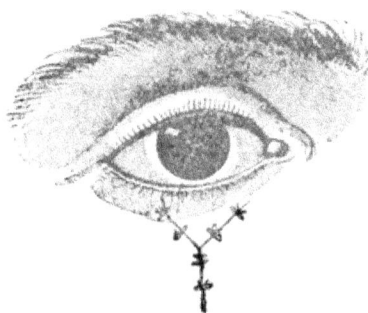

There are other methods, as Deffenbach's, Arlt's, Wolf's, etc., but as they are principally modifications of those already mentioned and more or less difficult to perform, the *modus operandi* of each is omitted.

The fact, in regard to all those cases requiring surgical treatment, is, that each method, no matter to what case it is applied, requires some modification; and the practical surgeon will vary from set rules and adopt such procedure as his mechanical ingenuity dictates.

ENTROPION.

Entropion is a condition of the eyelid characterized by a turning in, or an inversion of its edges against the ball.

Entropion is usually due to some organic

change in the structure of the lid, as cicatricial contractions of the conjunctiva or tarsal cartilage, or
to spasm of the orbicularis palpebrarum.

If the entropion is due to cicatricial changes
as before indicated, it is usually the result of a
trachomatous condition of the lid, and affects principally the upper lid.

When the entropion is due to spasm of the orbicularis muscle (spastic entropion), it usually affects the under lid, although it may affect both.
This condition is usual in old people who have had
an operation upon the eye, as for cataract, which
necessitates the use of a bandage for some time.
It also is seen in cases of severe blepharospasm and
photophobia, in children principally, as a result of
corneal ulcer.

Treatment.—Where the entropion is caused by
organic change in the lids and especially in the tarsus, an operation is usually necessary to correct the
defect.   I have, however, in some cases been able to
correct the defect by the continued use of collodion.
It should be applied, with a camel's hair brush, along
the edge of the lid and covering the entire portion
of the lid over the tarsal cartilage.   As soon as it
becomes detached, it can be reapplied.   I have kept
this treatment up indefinitely in both the spastic
and organic forms with much success.

In severe cases of the spastic form, no matter
from what cause, a canthotomy is the most to be
relied upon.   This is especially the case in the en-

tropion due to the blepharospasm and photophobia
as the result of corneal ulcer.

In the unyielding cases of the organic form of
entropion, there are several methods of operation,
owing to the condition of the parts requiring cor-
rections, which have been recommended, the most
practical of which are Arlt's Method and Streatfield's Operation, which are given below.

Arlt's Method as described by Juler:

"A small double-edged straight knife is inserted
at one or the other end of the eyelid between the
cilia and the meibomian ducts, and its point is made
to come out through the skin about two millimetres
above the lashes. It is then made to cut its way
along the whole edge of the lid, and thus forms a
bridge of tissue containing the lashes only. A second incision is now made from the two extremities
of the first, curving upwards to the extent of three
or four millimetres. This forms a semilunar flap on
the upper lid which must be dissected off. The
bridge of skin containing the cilia has now to be
shifted upwards, and its upper edge attached by
sutures to the skin of the lid, its lower edge being
left free. Simple water dressing is all that is necessary."

Streatfield's Operation:

"The lid is held with compressing forceps, the
flat blade passed under the lid, and the ring fixed
upon the skin so as to make it tense and expose the

edge of the lid. An incision with the scalpel is
made of the desired length, just through the skin,
along the palpebral margin, at the distance of a
line or less, so as to expose, but not to divide the
roots of the lashes; and then just beyond them the
incision is continued down to the cartilage (the extremities of this wound are inclined toward the edge
of the lid); a second incision, further from the palpebral margin, is made at once down to the cartilage in a similar direction to the first, and at the
distance of a line or more, and joining it at both
extremities; these two incisions are then continued
deeply into the cartilage in an oblique direction
toward each other. With a pair of forceps the strip
to be excised is seized, and detached with the
scalpel."

The remainder of the operation is described by
Juler as follows:

"Three sutures are then introduced as follows:
A small curved needle, armed with fine silk, is passed
first through the lower edge of the skin wound, then
through the upper edge of the groove in the tarsus,
and the two ends tied tightly together. The upper
edge of the skin wound is thus left free, and unites
very well without sutures."

TRICHIASIS.

Trichiasis is that condition in which the eye-
lashes turn in upon the ball, and irritate the eye.

As this condition is usually the result of an

ectropion, the treatment given under that head is sufficient.

DISTICHIASIS.

Distichiasis is that condition which indicates a superfluous or distorted growth of the eyelashes, in addition to the normal row, so arranged that they turn in upon the eye and irritate it.

Most all operations for this condition, outside that given for ectropion, have not proven eminently satisfactory, and have been generally rejected by the modern oculist, for the principal reason that the normal contour of the lids has been more or less interfered with.

In view of this fact, electrolysis is no doubt the very best method to be applied in these cases.

ANKYLOBLEPHARON.

Ankyloblepharon signifies an adhesion between the edges of the lids, and may be *partial* or *complete*.

In *partial* ankyloblepharon only a part of the edges of the lids are adhered.

In *complete* ankyloblepharon the entire edges of the lids are adhered.

Ankyloblepharon may be *congenital, traumatic* or the result of a *disease of the lids.*

Treatment.— The treatment, no matter what the cause, consists in dividing the adhesion with the knife or scissors. If a knife is used, the eye should be protected with a grooved director behind the parts to be divided, in order to fully protect the eye. A pair of scissors with a blunt point is the best instrument for this purpose.

It is a small matter to divide the adhesions in ankyloblepharon, but to keep them separated is quite another matter, for the tendency is to reunite. For this purpose the palpebral conjunctiva of one of the lids, preferably the lower one, should be stitched to the skin of the lid. It is often necessary to dissect the conjunctiva off the edge of the lid, and draw it out in order to meet the skin to which it is to be stitched. We must be careful that this procedure is well carried out at the angles of the lids.

It frequently occurs, generally as a result of burns, that the ankyloblepharon is complicated with a symblepharon, in which case we should endeavor to ascertain the extent of the latter prior to undertaking an operation, for if there are general adhesions between the lids and the globe, the operation will prove a failure.

The extent of the adhesion can be fairly well ascertained by watching the movements of the eye behind the lids, or better by passing a probe into an opening and ascertaining how far it can be introduced in all directions.

SYMBLEPHARON.

Symblepharon is an adhesion between the conjunctiva of the lids and the globe, and is *complete* or *partial*.

In *complete* symblepharon the palpebral conjunctiva is more or less united to that of the globe, and the cul-de-sac participates, to a greater or less extent, in the condition.

When the symblepharon is *partial*, the palpebral conjunctiva is united to the ball by larger or smaller bands, but leaving the cul-de-sac free.

Symblepharon, whether complete or partial, is usually the result of burns, and is more or less serious owing to the extent of the surface transformed into cicatricial tissue.

Treatment.— The treatment of symblepharon depends to a great degree upon the extent of the adhesions. If the adhesions are complete, involving the entire cul-de-sac, no operation should be undertaken.

When the adhesions are partial, the treatment depends upon their situation and extent. If there is but one or two small adhesive bands, we can often succeed in separating them by means of a ligature tied very tightly round the cicatricial tissue. If, however, the band is large, it can be divided into two, or even more parts, by ligatures surrounding each part, as in the small band.

In case of complete symblepharon, where only a part of the cul-de-sac is involved, the following operation recommended by Meyer is one of the best methods. It is, in his own words, as follows:

"The base of the symblepharon is pierced with a triangular needle, parallel with the palpebral fold,

and inserted as deeply as possible. Then a leaden
string is inserted in the wound made by the needle,
and its two ends are moulded so as to fit the angles
from which they emerge. Some surgeons unite the
ends of the thread, and, from time to time, tighten
the knot. The thread is left in situ till the wound
is cicatrized, when the adhesion is cut in the same
way as for incomplete symblepharon."

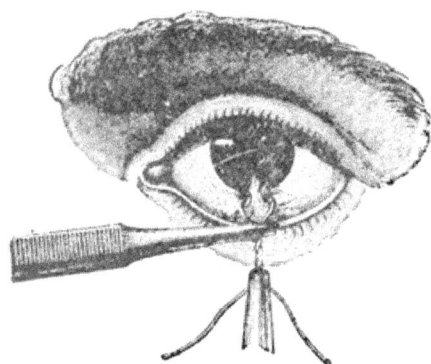

FIGURE 20. (After Meyer.)

BLEPHAROPHIMOSIS.

Blepharophimosis is that condition of the palpebral fissure in which it is contracted, and thereby
diminished in length.

In this condition the angles of the lids are
brought closer together than normal and is usually
the result of a trachoma.

Treatment.— Blepharophimosis is remedied only
by an operation known as canthoplasty. One of
the most simple and satisfactory operations for

canthoplasty is that of Meyer, and is described as follows:

"The external commissure is divided in its entire thickness in a line with the direction of the palpebral fissure. This section may be made with a bistoury, the point being gently inserted between the eyeball and the external commissure.

"The entire thickness of the integuments is then transfixed with the point of the knife from within outwards, and the whole commissure is easily divided by pushing the bistoury outwards."

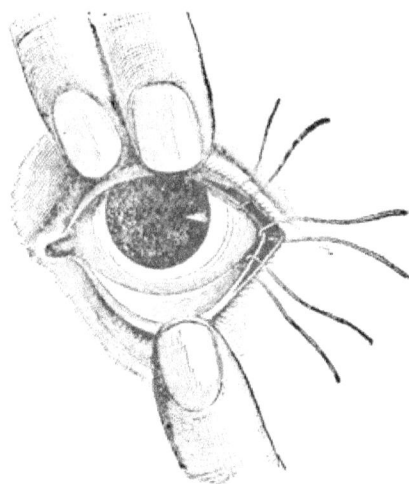

FIGURE 21. (After Meyer.)

"The operation is still more easily performed with straight scissors, one blade being introduced behind the commissure; the wound in the skin should always be a few millimetres longer than that in the conjunctiva.

The section of the commissure being thus completed, an assistant draws the margins of the wounds upwards and downwards, so as to change a horizontal into a vertical section. The surgeon takes hold of the conjunctiva near the centre of the section, and passes through it a very fine needle furnished with a silken thread; he then lets go the conjunctiva, and takes hold of the external skin also at the centre of the section; the needle is carried through the skin, and on tying the suture the corresponding margins of the skin and mucous membrane are brought together. In like manner two sutures are also inserted near the angles of the wounds."

EPICANTHUS.

Epicanthus is a congenital malformation of the lids, in which a fold of the integument at the inner canthus covers the caruncle to a greater or less extent. This condition is caused by superfluous integument over the bridge of the nose between the eyes.

Epicanthus can be improved by removing an oval piece of the redundant integument from the bridge of the nose, the amount removed being regulated by the extent of the deformity.

The margins of the wound must be so shaped as to insure perfect coaptation, and the edges must be held together by hair pin sutures.

CHAPTER V.

SECTION I.

DISEASES OF THE CONJUNCTIVA.

THE conjunctiva is subject to various affections, which may be designated as *inflammatory* and *non-inflammatory.*

By far the largest percent are inflammatory.

The following may be considered under the head of inflammatory diseases of the conjunctiva:

1st. *Hyperaemia of the conjunctiva.*

2nd. *Inflammation of the conjunctiva.* (Serous, Mucupurulent, Purulent, Plastic, and the Associated).

1st. Hyperaemia of the conjunctiva is that condition which is simply marked by an excess of blood in that structure.

Hyperaemia is *active* or *passive.*

Active hyperaemia is due to an increased inflow of blood.

Passive hyperaemia is due to an obstructed outflow of blood.

Hyperaemia of the conjunctiva is also *acute* or *chronic.*

The causes of hyperaemia of the conjunctiva are numerous, among which are: uncorrected ametropia, beginning presbyopia, incipient cataract, slight opacities of the cornea, local irritants, as foreign bodies, such as dust, wood or tobacco smoke, cold winds, the abuse of alcohol, and associated diseases, as nasal catarrh, lachrymal obstruction and blepharitis marginalis.

The point of transition between hyperæmia and inflammation is so subtile that it is impossible to determine just where hyperæmia ends, and where inflammation begins.

The symptoms of hyperæmia of the conjunctiva are slight congestion of the vessels, swelling of the conjunctival follicles, photophobia, lachrymation, and a hot stinging sensation.

Treatment.— Remove foreign body; correct refractive error (many cases of hyperæmia are caused by wearing glasses that are not properly centered, or otherwise not properly adapted to the eye); open obstructed tear duct; apply proper remedies to associated diseases.

Locally, R.—acid boracic, grs. v,
aqua destil, oz. i.
Mix and filter.

To be instilled into the eye three or four times a day in acute hyperæmia.

If the hyperæmia is chronic and caused from some of the associated diseases, as nasal catarrh, blepharitis marginalis, use the following:

R.—hydrastin, gr, ss
acid carbolic (pure) gtt, i
morphia sulph,

cocaine murias aa grs, iv.

glycerine drs ii.

aqua destil, drs vi.

First add the carbolic acid to the glycerine, and the other ingredients to the aqua destil; then unite the two solutions and filter. A few drops in the eye three or four times a day.

The advantage of this application is that the anaesthetic effects are very lasting, which is principally due to the carbolic acid. Always be sure that the carbolic acid is pure.

Besides the anaesthetic effects of carbolic acid, its antiseptic properties are not to be ignored in many affections of the eye.

If the hyperaemia is active, cold applications are indicated. If passive, hot fomentations, but never poultices in any affection of the eye. Raw beef, rotten apple, scraped potato, and all similar so called remedies, are only mentioned to be emphatically condemned.

2nd. *Conjunctivitis.*

Conjunctivitis is an inflammation of the conjunctiva *per se.* When the deeper structures become involved, the term *ophthalmia* should be applied.

The following division for the study of the different forms of conjunctivitis (the ophthalmias) is presented. This division being founded upon the character of the secretions, is considered the most logical arrangement.

CONJUNCTIVITIS, (Ophthalmia)	Serous	{ Acute catarrhal
	Muco-purulent	Chronic catarrhal Trachoma
	Purulent	{ Neonatorum { Blenorrhœal
	Plastic	{ Croupous { Diphtheritic

SEROUS CONJUNCTIVITIS.

Serous conjunctivitis is characterized by simply a watery discharge from the eyes. There is inflammation of the palpebral, fornix and ocular conjunctivæ, characterized by congestion and redness of these parts. There is usually a sensation as if sand or other foreign body were in the eye. There may or may not be photophobia. As long as the discharge is serous (watery), it is classed as acute catarrhal, and the discharge is not considered contagious. As soon, however, as mucus or muco-purulent matter is found in the discharge, it ceases to be acute catarrhal, but merges into the muco-purulent variety, and the discharge from the eye is regarded as contagious.

The causes of serous conjunctivitis are usually due to climatic changes (colds), foreign bodies in the eye, irritating substances in the atmosphere, as dust, smoke, etc. Eye strain is also a frequent cause.

Treatment.— Examine the eye carefully to ascertain the presence of foreign body. A hair from the lid, a cinder or other foreign body is very liable to fall upon the ball, when, if it does not adhere (most always to the corneal surface), it is either swept away by the lachrymal secretions, or is carried by the action of the lids to the under surface of either lid, mostly, however, to the under surface of the upper lid.

In all cases of conjunctivitis, whether there is a history of contagion or not, examine the eye thoroughly for foreign body:

First: Cocainize the eye by instilling a few drops of a 4 per cent solution on its surface, or between the lower lid and the ball. Then inspect the cornea carefully, viewing it from several directions. It frequently occurs that a foreign body embedded on the surface of the cornea is not seen because we do not observe the point from the proper direction. Then evert the upper lid; generally if there is a foreign body in the eye, it will be found adhering to the palpebral surface of the upper lid.

As this procedure will not give you a full view of the entire conjunctival surface, insert some smooth instrument under the everted lid, and at the same time direct the patient to look down. Raising the lid with the instrument, you are enabled to inspect the entire surface of the superior fornix.

This is important, for it occasionally occurs that a foreign body drifts into the superior fornix. I once removed a stem of ragweed, to which was attached two or three of the seed, that had remained in this part of the eye for more than six months.

The surface of the inferior fornix can be brought to view by making slight pressure with the finger upon the integument of the lower lid and pressing it toward the cheek bone, at the same time directing the patient to look upward.

If no foreign body is found and there is no history of irritation from dust, smoke, or extraneous causes, then we can attribute the condition to climatic changes, or to eye strain due to refractive irregularities, or to overwork.

To abort the inflammatory condition, the following solution should be applied to the conjunctival sac once a day:

R.—argenti nitras (cryst.) gr i
    aqua destil. oz i
    Mix.

Invert the upper lid and apply to its conjunctival surface with a small cotton applicator, well saturated with the solution. At the same time if the lower lid is slightly pulled away from the ball, the inferior fornix will receive a portion of the solution—if immediately after the upper lid is returned to its place, it is gently lifted from the ball by the cilia, the solution will come in contact with all parts of the conjunctival sac.

Together with the above, instil a few drops of F. No. 2 into the affected eye three or four times a

day or oftener if necessary. I instruct my patients
to apply the solution whenever the eye becomes
painful.

If the lids become matted together in the morning, apply to their edges upon retiring at bed-time
a very small portion of vaseline, just enough to moisten the roots of the cilia.

Frequent attacks of serous conjunctivitis is indicative of eye strain and the eyes should be thoroughly examined in that direction as soon as the inflammation subsides.

There are many other acceptable remedies for
serous conjunctivitis besides those which have
already been mentioned, but avoid the sulphate of
zinc, sugar of lead, sulphate of copper and other
strong caustics, as their use in any case is of doubtful benefit, but is often fraught with positive injury.

MUCO-PURULENT CONJUCTIVITIS.

Muco-purulent conjunctivitis is characterized
by more or less chronic inflammatory condition of
the conjunctiva. There is loss of lustre of the palpebral conjunctiva together with congestion of both
the palpebral and ocular conjunctivae.

There is usually present photophobia and often,
although not always, blepharospasm. There is a
serous and muco-purulent discharge, and the lids are
glued together on awakening in the morning. There
is a sensation as of sand or other gritty substance in

the eye, caused by the presence of mucus and the enlarged papillae.

In severe cases the lids become thickened, there is blepharitis, there is increased conjunctival irritation caused by closure of the punctum, or its eversion from the ball, and the resulting epiphora.

In severe and protracted cases of muco-purulent conjunctivitis, there are liable to be corneal complications, such as pannus and corneal ulcer. In case of either, the photophobia and blepharospasm are usually severe.

As a neglected hyperaemia or a serous conjunctivitis may develop into the muco-purulent form, the causes are practically the same with the exception that a muco-purulent conjunctivitis is liable to be produced by a contagion.

Here is a matter of great importance: micro-organisms are present in every case of muco-purulent conjunctivitis, no matter what its cause, or whether it is severe or not, and are liable if they come into contact with an eye to set up not only a muco-purulent ophthalmia, but a purulent form of the most severe type.

The various terms, "granular lids," "granular ophthalmia," "papillary granulations," "follicular granulations," "trachoma," etc., are in fact synonymous and doubtless only different forms of muco-purulent ophthalmia.

Granulations are not to be regarded as pathological formations, but simply changes in the lymph

follicles, and the terms "papillary trachoma" and
"follicular trachoma" are but different conditions of
the lymph follicles, owing perhaps to the different
grades of inflammation of the conjunctiva. It has
been well demonstrated that "granulations" of the
lids differ from the pathological granulation-tissue
in that the mucous membrane in granulated lids is
not ulcerated.

Juler claims that the granulations are eleva-
tions composed of aggregations of lymphoid cells,
beneath the mucous membrane, with a partial fatty
degeneration of these cells nearest the surface.

Although muco-purulent ophthalmia, from its
contagium or otherwise, may attack anyone, yet certain individuals are predisposed to it, especially
those living in poorly ventilated rooms, and those
whose nutrition is enfeebled by syphilis or tuberculosis.

Muco-purulent ophthalmia is extremely contagious, and on account of the diverse ways in which
its contagium is spread, it is almost sure to attack
every inmate, when it becomes developed in crowded
institutions, such as those for destitute children.
These children, as a rule, are not of vigorous constitution; many of them being of scrofulous or tubercu-
lous habit are very liable to become affected with
contagious diseases in general.

Muco-purulent ophthalmia is so unrestricted in
its extent both as it regards the field it invades and
the character of the attendant inflammation that

scarcely two cases are similar except in a few general features, therefore the impossibility as well as impracticability of anything like a specific medication.

The palpebral surface of the conjunctiva may alone become affected, as it usually is the point of attack in the incipiency of the affection; or the entire conjunctival sac may become involved.

Not only may the conjunctiva suffer, but the deeper structures often become seriously implicated.

A clear distinction should be made between muco-purulent conjunctivitis and muco-purulent ophthalmia. When the affection is confined to the conjunctiva alone, without the involvement of other adjacent structures, then the affection is a *conjunctivitis*, but when other structures become involved, as the lymph glands and follicles, and the various secretory and excretory ducts, then the affection should be known as an *ophthalmia*.

Treatment.—The first requisite and indispensable in the treatment of muco-purulent ophthalmia is a most thorough and pains-taking cleanliness. This is essential, not only for the good of the patient, but in order that the affection may not be communicated to others.

As there is a condition of sepsis, antiseptics are always indicated in advance of, as well as in connection with other treatment.

The kind of antiseptic is also a matter of inportance, in eye affections especially.

Bi-chloride of mercury, both as a germicide and as an antiseptic, heads the list, and the solution of 1-5000 is indicated in muco-purulent ophthalmia of the chronic non-inflammatory variety. I have known cases of this kind improve rapidly under its use, and many such cases I believe require no other treatment.

But where there is much inflammatory action, as in complications, as of pannus or corneal ulcer, then the bi-chloride of mercury is contra-indicated. Its use in such cases very frequently increases the inflammation to a very exasperating degree.

Carbolic acid as an antiseptic fills an important place in the treatment of diseases of the eye, not for its antiseptic qualities alone, but for its anesthetic effects. Where there is much inflammation, the carbolic acid solution (2 to 4 per cent) is indicated. The soothing influence which the solution has in these cases, especially where complicated with corneal ulcer, is remarkable. Its effects in many of these cases is as prompt as cocaine and much more lasting.

When it is determined which antiseptic is to be used, all of the secretions should be removed with pledgets of absorbent cotton, after which the lids should be thoroughly washed with sterilized water. The antiseptic should be so applied that it will come in contact with all parts of the conjunctival sac.

This can be done by first applying it from a dropper to the conjunctival surface of the everted upper lid; next, the lower lid should be pulled slightly away from the ball, and the solution from the dropper, as before, is instilled into the lower fornix, filling it. Then by slightly everting the lower lid, and at the same time tilting the everted upper lid away from the ball, by a slight pressure upon its edge, the lower lid thus everted can be dextrously pushed up under the upper lid, and will carry with it the solution to all parts of the conjunctiva.

PURULENT OPHTHALMIA.

Purulent ophthalmia is an acute inflammatory condition of the conjunctiva, characterized by a high state of inflammation and a profuse purulent or muco-purulent discharge from the eyes, caused by inoculation from a specific contagium.

This condition is also known as acute blennorrhoea, which signifies a profuse discharge of mucus, but purulent ophthalmia is considered a better term, as the discharge consequent to the affection partakes of that character.

Purulent ophthalmia is considered under two heads, viz. *gonorrhoeal ophthalmia* and *ophthalmia neonatorum*.

Gonorrhoeal ophthalmia is caused from infection with the gonococci, the specific contagium, or morbid principle of which incites the disease.

The means of conveyance of the poison into the conjunctival sac in gonorrhoeal ophthalmia are too numerous to mention; the most common, however,

being from uncleanliness, in not keeping the hands
well cleansed after manipulation of the affected
parts.

The contagium can also be carried to the eye by
washing in the same basin and wiping upon the nap-
kin or towel used by one who has the affection.   This
is a most prolific manner of spreading the disease.

The contagium can also be carried with the
secretions from the eye to the floor or the street, and
after drying may be conveyed to the eye with the dust
raised in the sweeping of the floor, or by the wind
on the street.

In ophthalmia neonatorum there are not so
many ways in which the eyes become affected, as it
occurs during or after the child's head passes
through the vagina, or from inoculation after birth,
through negligence of the nurse or mother by con-
veying the virus from her fingers, or through the
towels while bathing the child.

It is a question whether all cases of ophthalmia
neonatorum are not caused from a vaginal discharge
that is gonorrhœal; and while such is probably the
case in the severer forms, yet it is not sympathetic
nor charitable to attribute this cause to all cases.

I have seen very severe cases of ophthalmia
neonatorum in which the gonococcus could not be
discovered in the conjunctival discharge, and which
affection was due to a vaginal discharge which was
not of the specific gonorrhœal character.

We must also remember in giving our opinions

as to the cause, that the discharge may be of gonorrhoeal origin, and the mother be entirely innocent
and blameless in the contraction of the affection.   It
is therefore the province of the physician to relieve
his patient and prevent if possible impending blindness, rather than speculate as to the origin of the
affection.

Purulent ophthalmia, whether in the infant or
adult, is a highly inflammable and contagious affection, caused by infection, making its appearance in
from two to four days after the virus has been
brought into contact with the conjunctiva.

The symptoms of both forms of purulent ophthalmia are so similar that it is not necessary to
treat of them separately.

The first symptom of a purulent ophthalmia is
a serous discharge from the eye, soon followed by a
serous infiltration of the palpebral and ocular conjunctiva.

The infiltration of the ocular conjunctiva becomes so great that the conjunctiva is forced outward between the lids to such an extent that it is
impossible, often, to evert them, and in some cases
the swelling is so great that the lids are inverted,
and the cilia are turned in upon the conjunctiva.
This condition of the conjunctiva is known as *chemosis.*

This stage lasts from two to four days, during
which time the serous discharge changes to the muco-
purulent, then to the purulent, in its most virulent
and contagious form.

The chemosis now gradually decreases, but the lids continue to swell, the upper one becoming so large and pendulous that it hangs over the under lid, whose cilia add to the irritation by rubbing against the conjunctiva of the ball and the upper lid.

The chemosis and the swelling of the lids, together with the existing conjunctivitis now, by degrees, slowly subside, the discharge becomes less, and in the course of about four weeks the eye is usually left in its normal condition.

The affection usually attacks one eye, but in a short time the other becomes affected from a transmission of the contagium from its fellow. This is a typical course of an uncomplicated case of purulent ophthalmia.

But the eye frequently suffers from complications which imperil its safety to a serious extent, the most important of which is the dense chemosis of the ocular conjunctiva. This condition endangers the safety of the eye, because the nutrition of the cornea is obstructed by it. Then follow corneal ulcer, sloughing of the whole cornea, and its concomitant, staphyloma, with entire loss of vision.

Although corneal complications are the most to be dreaded, there are others, from the extension of the inflammatory condition, the most important of which is iritis, and often cyclitis and choroiditis.

Treatment.—The treatment of the purulent ophthalmias consists, from the onset, in thorough asepsis, in all that the term implies.

This is a most important matter, for it not only concerns the patient, but all others in the household. I have known all the members of a large family, where no particular concern was given to cleanliness, to contract the affection, coming from a single individual. This precaution is necessary whatever the type.

In case of the infant, especially, if we have reason to suspect an existing gonorrhoea, or even an abnormal vaginal discharge in the mother, the eyes should be thoroughly disinfected immediately after birth, with a 1-5000 bichloride of mercury solution.

This I consider much better than Crede's method of applying a 10 gr. solution of the argenti nitras to the conjunctiva, as a prophylactic; in fact the 10 gr. solution of argenti nitras is too strong for the eyes of an infant, from the fact that the epithelium covering the cornea is very delicate,—so very delicate, indeed, that the above solution, if allowed to come into contact directly with it, would abrade it to a considerable extent.

In order to thoroughly disinfect the whole of the conjunctival sac, we should use a small bulb syringe with a flattened nozzle which can be easily intruded between the lid and the ball.

With this instrument we can thoroughly flush and cleanse the conjunctival sac by gently raising the lid and intruding the nozzle between the lid and the ball, and pushing it as far up into the fornix as possible before expelling its contents, at the same time being careful not to exert any pressure on the ball with the point of the instrument, but instead, toward the lid. This is very important, especially in case of corneal ulcer, for the least pressure in such case is often very injurious to the eye. More trouble is experienced in flushing the upper cul-de-sac than the lower.

As a prophylactic, one flushing, immediately after birth, is quite sufficient; but if the affection has already set in, then the flushings should be kept up at least twice a day.

As fast as the purulent discharge accumulates it should be removed with pledgets of absorbent cotton, and a 10 gr. solution of boracic acid should be instilled, by means of a dropper, between the lids, as often as every two hours.

If there is much chemosis and swelling of the lids, the application of the solution is not of much importance, from the fact that it comes into contact with but a very small portion of the conjunctival sac.

The greatest danger to the eye, as heretofore indicated, is the chemosis, especially when it is so great as to interfere with the nutrition of the cornea.

If the chemosis is very dense and is protruded between the lids, and is reflected over the cornea to a greater or less extent, and the lids are swollen so much that they have become inverted, and their edges are burrowed deeply into the infiltrated conjunctiva, the best thing to be done is a free canthotomy, after which warm water fomentations, as hot as can be borne, should be applied to the eyes.

The fomentations should be changed as often as

every five minutes, and replaced with hot ones, which process should be kept up as long as an hour at a time, twice or three times a day.

Although nitrate of silver has been highly recommended in the purulent ophthalmias, I consider its use as highly dangerous, especially with infants.

The structure of the eye of a child a few days old is very delicate, the sclerotic, which is considered the most substantial part, being so thin that the choroid can be seen through it.

The cornea, as we all know, is as thin and delicate in in its structure as the sclerotic, but as its clearness is an indispensable to good vision, it is less able to withstand strong solutions, especially when of doubtful utility.

Many authors have recommended as much as a 20 gr. and some even a 30 gr. solution of the nitrate of silver in the purulent ophthalmia of infants. Now while the eye of the adult may be able to withstand these strong solutions, I am very sure that they are positively injurious in children, for if the use of a 10 gr. solution is kept up for even a very few days, the cornea will frequently soften and we will be surprised at one of our visits to find that it has given way, and that the iris is protruded through it.

Such an occurrence is usually attributed to the disease, when in fact it is the result of the nitrate of silver upon the cornea.

It should be remembered that atropine has no mydriatic effect upon the eye of an infant, and for this

reason it is unnecessary to use it, even if there is
an indication for the pupil to be dilated.

My own experience is that the flushings of the
conjunctival sac with the bichloride of mercury solution, 1-5000, is the most satisfactory of all known
medications in the purulent ophthalmias of the adult
and infant.

PLASTIC OPHTHALMIA.

The plastic forms of ophthalmia are found in
the *diptheritic* or membranous, and the *croupous*
or psuedo-membranous.

Plastic ophthalmia is, fortunately, not a frequent
affection in this country; however, occasionally we
have to deal with a case; but Germany and many
of the older countries have suffered very much from
its effects.

Although plastic ophthalmia is said to be
directly traceable to diphtheria, yet cases have been
known to exist where no trace of diphtheria could
be found, except that upon the conjunctiva.

The membrane in plastic ophthalmia is an exudation, caused by the coagulation of fibrin.

In the *diphtheritic* form the coagulation of fibrin
takes place in the whole thickness of the conjunctiva, down to the sub-mucous tissues. This is the
most dangerous form of plastic ophthalmia, and seldom leaves a patient without more or less permanent
loss of vision.

In the *croupous* form the exudation is poured out upon the surface of the conjunctiva, forming a pseudo-membrane, which can be lifted off of the conjunctiva, very readily, with forceps, without injury to it. This form of plastic ophthalmia is of no great consequence, as it never amounts to more than a catarrhal conjunctivitis.

The prominent symptoms of diphtheritic ophthalmia are extreme tension and swelling of the eyelids, accompanied by a sense of heat, and often, although not always, severe pain. The lids are so stiffened by the exudate that it is almost impossible to evert them. This condition of the lids is one of the leading features in the diagnosis of this affection.

Diphtheritic ophthalmia is a serious affection, often destroying the eye within a day after its inception. In some severe forms, treatment is seemingly incapable of the least benefit.

The greatest danger in diphtheritic ophthalmia is that the exudate becomes so impacted into the conjunctival and sub-conjunctival tissues that the nutrition is interfered with to such an extent as to cause ulceration and necrosis of the cornea.

All cases of diphtheritic ophthalmia, in the course of from thirty-six to forty-eight hours, gradually pass from the plastic stage to that of a purulent character. This is characteristic of the affection, and is accompanied by a return of the flexibility of the lid, in that it is now more easily everted, and by the purulent character of the discharge.

Treatment.—The treatment in the onset, and during the plastic stage, consists simply of iced cloths to the lids, and the flushing of the eye with the bichloride of mercury solution, 1-5000.

As soon as the purulent stage sets in, the iced cloths must be changed for hot water fomentations, the same as in the purulent ophthalmias.

As a prophylactic, in case the cornea should become involved, the pupil must be kept widely dilated with a 1 per cent solution of atropine, which should be instilled into the eye three or four times a day.

It will be very fortunate if the cornea does not become involved before the suppuration begins, as it rarely occurs afterwards, but the tendency is usually to a speedy recovery, with the treatment already laid down for the purulent ophthalmias.

PTERYGIUM.

Pterygium is a triangular fold of mucous membrane, growing out from the conjunctiva of the eyeball, over the cornea.

That part of the pterygium at the apex of the triangle, and attached to the cornea, is named the *head;* its base, or that part at the canthus of the eye, is termed the *body;* and that portion immediately behind the apex, at the sclero-corneal border, is denominated the *neck.*

There are many opinions with regard to the origin of pterygium, but the consensus of all is that the starting point is usually a pingueula, or else an ulcer situated near the sclero-corneal border, on the conjunctiva.

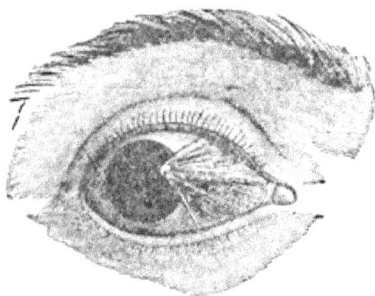

FIGURE 22. (After Meyer.)

Pterygium consists of hypertrophied tissue, often very trifling in amount, but sometimes very considerable and unsightly. This morbid growth is fibrous in structure and is not firmly adherent to the cornea and sclerotic. It is covered with conjunctival tissue, which surrounds the hypertrophied mass, which is entirely disconnected from the conjunctiva, except at the linear adhesion on its under surface, where it acts in the manner of a pedicle.

It seldom reaches or passes the center of the cornea, for the reason, perhaps, that the blood vessels which feed it do not extend further, usually, than that point.

Although the affection is of little importance, unless it implicates the cornea to such an extent as to obstruct the vision, yet its appearance to many, and especially to its possessor, is more or less repug-

nant, and patients are generally solicitous to rid
themselves of its presence.

Without attempting to discuss the cause of this
morbid growth, it is our object to consider the most
feasible means for its thorough and permanent removal.

Authors mention various methods for this purpose, which resolve themselves into three plans of
operating, viz.: excision, transplantation and ligation, or a combination of two or all of these methods;
thus, one recommends its entire removal by excision,
by dissecting it from its apex to the semilunar fold,
where it is excised; another recommends dissection
of the apex from the cornea to its base, and then
inserting the dissected portion underneath the conjunctiva, where it is held by sutures; another advo-
cates the ligation of the base and the dissection and
excision of the apex; and finally, another the ligation of both the base and apex together with the
narrow strip underneath the growth, where the ligatures are allowed to remain until the growth sloughs
away.

If any of these methods or any of their combi-
nations have given general satisfaction, I am not
aware of it; but, on the contrary, oculists usually
discourage patients from having anything done with
the affection unless it encroaches so far over the cornea as to interfere materially with vision.

In a practice of more than thirty years, during
which time I have attempted the removal of quite

a number of these growths by one or another of the methods herein enumerated, I cannot recall more than a half dozen cases in which the result was all that could be desired; but, on the other hand, many of them were not benefited in the least, the operation frequently having resulted in unsightly cicatrices.

Arlt has demonstrated that pterygium enters into the substance of the cornea, beneath its epithelial layer, yet, as previously mentioned, the connection is not firm, and its fibres can be readily separated from the cornea with very little effort.

A few years ago, before the anaesthetic properties of cocaine were discovered, I attempted to remove a large pterygium in the case of a man who was somewhat under the influence of liquor. I caught the growth near its apex with the forceps. When I was about to separate it from the cornea with the scissors, he made a desperate effort at resistance, and caught my arm with which I was holding the forceps so suddenly that I did not have time to relax my hold. It bled freely, and I was fearful that the eye was badly injured. After a considerable time I was allowed to cleanse the eye, when I found that the pterygium was entirely separated from the cornea and as far back as midway between the sclero-corneal junction and the caruncula lachrymalis.

That part of the cornea to which the growth had been attached was somewhat hazy, and presented the appearance of numerous small depressions not

larger than pin-points, showing the points where
the fibres had penetrated the cornea. The patient
would not permit me to proceed farther—not even
to excise the loose tissue remaining.

I recommended the application of cold water to
the eye and advised him to return daily, that I might
watch the eye. He went home and, as he suffered
no pain, concluded he was doing well enough, and
did not return for three weeks, at which time the
detached tissue had contracted and I could scarcely
see any corneal opacity or other evidence that the
eye had ever been affected with pterygium.

What I so much feared in this case was that
the corneal layers had been separated and, as a result, corneal ulceration; or, on the other hand, infiltration between the separated layers and a consequent extensive opacity.

Subsequently I thought little of the occurrence
more than to congratulate myself how fortunate I
had been in not having destroyed the man's eye,
until some time afterwards I was relating the circumstances to a physician, when he gave me the following:

Aa old gentleman of his acquaintance had a
large pterygium on each eye. He had been having
his eyes "treated" from time to time by different
parties, principally traveling "specialists," without
benefit, when, in conversation with his family physician, he was informed that treatment in such cases
was not generally successful, that his eyes would

probably not become worse, and he was advised to desist.

The old gentleman had his own idea about the matter, and insisted that the growth would continue to "cover the sight" and in a short time leave him hopelessly blind. He argued that his disease was similar to that of the horse, known as "hooks," and that "hooks" are cured simply by "pulling them off." He concluded to act in his own behalf and, having procured a small pair of forceps, such as is commonly known as "eye tweezers," he actually removed the pterygia from both eyes, and without subsequent treatment the procedure effected a permanent cure.

I have ascertained that the affection known as "*hooks*" in the horse is simply an attachment of the membrana nictitans to the cornea, the adhesion being, probably, the result of an inflammatory action of this membrane and the ocular conjunctiva.

Shortly after this, I presume about two years ago, I noticed in some medical journal that a foreign oculist, Arlt, I believe, was removing pterygium by evulsion, and being greatly interested, by reason of the circumstances just mentioned, I have been anxiously waiting to hear with what success. So far I have been disappointed.

For the past ten years I have rarely used other means in the removal of pterygium than evulsion. I have not had a great number of cases, but the results have been generally satisfactory.

In my first operations I removed the apex, and as far back as the caruncula lachrymalis, with the forceps, and excised the loose tissue with the scissors. Afterwards I ligated the base first and then removed the growth with the forceps as far as the ligature, and excised as near it as possible. In this I found a great advantage. inasmuch as the ligature prevented haemorrhage and I could operate to a better advantage.

Recently I have discarded the forceps, and taken in its stead a blunt hook—such as is used in the operation for strabismus—which I find a much better instrument for this purpose.

After the base has been ligated the pterygium is separated from its attachment with the blunt hook, by running it under a small portion at a time, especially if the pterygium is large; first separating the thin connection between the ligature and the cornea; then a small portion of the corneal attachment at a time, until the whole is removed.

The loose tissue is excised as close to the ligature as possible, cold water dressings are used, and the ligature is allowed to remain until the strangulated portion sloughs, when it, with the ligature, will become detached and pass off.

Instead of using the ligature, I occasionally excise the pterygium near its base, after which I entirely cover the wound by drawing the conjunctiva together with sutures.

The advantage of this operation is the complete and thorough removal of the growth from the parts into which it has been imbedded, without the injury or superfluous removal of any normal tissue.

CHAPTER VI.

SECTION I.

DISEASES OF THE CORNEA.

THE term *corncitis* signifies an inflammation of the cornea. *Keratitis* is a synonymous term and is generally applied to the different forms of inflammations of the cornea.

Keratitis
- Interstitial (diffuse)
  - Parenchymatous
  - Syphilitic
  - Strumous
- Punctate.
- Vascular (Pannus)
- Ulcerative
  - Primary
  - Secondary
  - Systemic
- Suppurative
  - Diffuse
  - Circumscribed

INTERSTITIAL KERATITIS.

In *interstitial keratitis* the entire cornea gradually assumes a chronic inflammatory condition without an inclination to the formation of pus or to ulceration. There is first a diffused grayish opacity of the cornea at the center, at first very slight, afterward very opaque, which gradually extends over the entire cornea, giving it a ground glass or steamy appearance, in which condition the iris and pupil are greatly obscured.

In very severe cases of instertitial keratitis the cornea presents a yellowish appearance, completely obscuring the iris and pupil.

The degree of inflammation, photophobia, ble-pharospasm, lachrymation and pain varies, not with the amount of corneal opacity, but from a peculiar hypersensitiveness not explainable.

There is a kind of pannus sometimes present in this affection which does not always come from the conjunctival blood vessels which run over the super-ficial surface of the cornea, but comes from branches of the ciliary vessels deep down in the corneal tissue. They are dullish red and named "salmon patches."

Frequently the iris and ciliary body become im-plicated with the disease.

This affection usually attacks both eyes, how-ever rarely at the same time, the interval being variable, from a week or more to months, or even a year.

It occurs between the ages of three and fifteen usually, although it may be seen earlier, and again it may occur much later.

The duration of the affection is usually from four months to a year, or even longer, under the very best directed treatment.

The effects of the disease are usually lasting, and distinct vision is scarcely ever attained.

Inherited syphilis is the usual cause, although in some cases a specific history is wanting, and a generous diagnosis would place the cause to "want of vitality" or "lack of proper nourishment."

This affection is usually found in anæmic and weak persons, and the general health and hygienic conditions must first of all receive attention.

Treatment.—The medical treatment consists in a protracted course of alteratives, the principal of which is the iodide of potassium.

The iodide of potassium should always be prescribed in the saturated form and given in milk. Iodide of potassium given in this manner never irritates the stomach, and children from two to ten years of age can easily tolerate fifteen grains three times a day.

If for any reason the iodide of potassium cannot be tolerated, then inunctions of mercurial ointment in the groins and under the arm pits once a day until the mouth or gums begin to show its effects, when it should be discontinued until its effects have passed off.

The syrup of the iodide of iron, and the hypophosphites of iron and strychnia, may be prescribed in the absence of the iodide of potassium, but there is no remedy so far as I know that acts so promptly as the iodide of potassium.

I always give it in three or four tablespoonfuls of milk, and before meals. For adults I prescribe from one-half to one teaspoonful (30 to 60 grains) in milk, as for children.

During the active inflammatory stage, a few

drops of a one per cent solution of atropine should
be instilled into the eye two or three times a day
in order to prevent adhesion of the iris to the lens
capsule.

The yellow oxide of mercury ointment in varying strength, usually from one-half grain to one
grain to the dram of vaseline, has always been recommended as a local remedy to the cornea in interstitial
keratitis, but the difficulty in obtaining the preparation free from irritating substances is almost impossible, and I rarely, for that reason, prescribe it.

Flushing the eye once or twice a day with a
1-5000 bi-chloride of mercury solution is much
cleaner and reaches the affected part much better
than is possible with the ointment.

The fact is, no ointment should be applied to the
conjunctival sac, as its ingredients scarcely ever
come in contact with the affected part, for the reason
that the vehicle is oleaginous, and is immediately
carried off, with whatever it contains, by the secretions of the eye before it could possibly have very
much, if any, effect.

As children mostly suffer from this affection,
they are inclined often to romp and play to fatigue.
While exercise and fresh air are necessary, overheating should be avoided.

Good, healthy, nutritious food is essential.

KERATITIS PUNCTATA.

Keratitis punctata is characterized by small spots on the posterior surface of the cornea, due to exudation of lymph from the iris and ciliary body.

It is not a disease by itself considered, but the effects of an inflammation of the iris, ciliary body, or choroid.

Keratitis punctata is a concomitant of serous choroiditis.

It occurs very often in the sympathizing eye in sympathetic ophthalmia.

It is usually of triangular form, the apex being near the pupil and the base below in the corneal periphery.

The only part of the cornea that becomes affected in keratitis punctata is Descemet's membrane, and this, because of the irritating properties of the exudate which is thrown from the iris and ciliary body into the aqueous humor.

Treatment.—As specific causes are almost certain, and inflammation of the entire uveal tract, constitutional remedies are indicated.

The general health in all cases must be improved.

No local treatment is necessary unless the eye is seen during the exudation of the lymph and while there is inflammatory action; then the pupil must be kept dilated with a one per cent solution of atropine.

VASCULAR KERATITIS.

Vascular keratitis is marked by a superficial vascularity and consequent opacity of the cornea.

The vessels causing this vascularity, which in the normal state are invisible, become so large as to be plainly seen with the naked eye.

The vessels thus making a pannus are continuous with those of the conjunctiva.

When there are few of these vessels, and the opacity is not so great as to prevent their being seen separately, they are found to be very tortuous, much more so, than those which are met with in other inflammatory conditions of the cornea.

The cause of pannus is generally a granular conjunctivitis, whereby the cornea is rubbed and thus inflamed by the diseased lids. It is also frequently caused by an entropion.

In pannus there is always present pain, impairment of vision, photophobia and lachrymation. Sometimes the inflammatory action is very great; at other times it is slight.

So with the pain; sometimes it is very severe, at other times it is almost absent, or simply an "uncomfortable feeling."

The most dreaded complication is corneal ulcer, to which this condition of the cornea is very much inclined; should an ulcer appear, it is very apt to run a severe course, and endanger the eye, because of the vascularity of the cornea.

The *treatment* of pannus is to ascertain the cause and remove it.

If from granulated lids, treat that affection in the most approved manner. If from entropion, an operation and removal of the hairs.

Peritomy is not practical, and pus and jequirity inoculations need only be mentioned, to be emphatically condemned.

ULCERATIVE CORNEITIS.

A *corneal ulcer* is a morbid disintegration of the corneal tissue, attended by more or less inflammatory action.

There is an endless variety of corneal ulcers, but for convenience of study corneal ulcers may be classified, according to their etiology, into the *primary*, the *secondary*, and the *systemic*.

Primary ulcers are usually confined to one spot, and are most frequently caused from some direct injury to the cornea, as an abrasion from a foreign body, or they may be due to a disturbance in the nutrition of the cornea, as in glaucoma, and in purulent conjunctivitis, where the chemosis interferes with the lymphatic circulation.

Primary ulcers may be superficial or deep, according as they affect the superficial structures alone, or burrow deeply into the corneal tissue.

Secondary ulcer of the cornea has its starting point in some other structure of the eye, most frequently the conjunctiva, and passes over to the cornea.

Secondary ulcers may also be superficial or deep. The *serpiginous* and *herpetic* forms belong to this class.

Systemic ulcer is due to some affection of the system at large, as scrofula, tuberculosis, or defective nutrition.

The most common form of this is the *phlyctenular*. Another less frequent form is the *ulcus serpens.*

Subjectively, the first indications of a corneal ulcer are more or less pain, with a sensation of a foreign body within the eye, an inflammation of the conjunctiva and lachrymation.

Objectively, upon examination is found an opaque spot upon the cornea, the surface over which is somewhat raised and cloudy; this spot marks an infiltration in the corneal tissue.

After the infiltration is formed the epithelium soon gives way over the spot, and is followed by a loss of substance of the corneal tissue; we have now a corneal ulcer.

The first appearance of a corneal ulcer after the rupture of the epithelium, and breaking down of the corneal tissue, is the clouded or gray appearance of the walls of the ulcer. This condition is caused by the remaining infiltrate.

If the infiltrate is thrown off, and the ulcer is soon cleansed, and the tendency of the ulcer is to heal at once, we have now what is regarded as a *superficial ulcer.*

But if the infiltration extends *en masse*, or in the form of slender stria into the body of the cornea, barrowing downwards into the corneal substance, it may distinguished as a *deep ulcer*.

But if the infiltration extends in area, with well defined edges, or in the form of slender stria in different directions in the transparent cornea, it is distinguished as a *progressive ulcer*.

With these considerations we are now prepared to take up the subject of *Corneal Ulcers* in regular order.

Primary Ulcer.—As indicated heretofore, a primary ulcer is one which originates in the cornea itself, and is usually the result of traumatism.

As the cornea is covered by epithelium, this covering always suffers in corneal ulcer, and frequently it is the only tissue originally involved, for the most formidable ulcers are often the results of the slightest abrasions of the epithelium.

I cannot, however, believe that the wounding of the corneal epithelium alone is sufficient to cause ulceration without an underlying cause, such as the introduction of extraneous matter of irritating or septic tendencies, for large spaces of epithelium are often removed from the cornea, which is replaced in the remarkably short period of twenty-four hours.

Another contingency in primary ulcer which has an important significance, is the age and nutritive condition of the patient.

In youth, corneal ulcers of the primary type

are almost unknown, except where there are pre-
disposing proclivities, as in scrofula or defective
nutrition. In such conditions I have known a very
slight abrasion of the cornea to result in a very
destructive ulcer.

A conjunctivitis, nasal catarrh, or lachrymal
abscess aggravates corneal ulcer, and, no doubt, in
many cases of slight abrasion of the cornea, pro-
vokes severe inflammatory action from contact of
the septic exudations consequent in these affections.

In the aged a very slight corneal abrasion often
results in opacity of the entire cornea. This is generally due to defective nutrition, although quite
often to the discharge from a lachrymal abscess.

The discharge from a lachrymal abscess is of
pyogenous character, and is liable to be filled with
micro-organisms, whose presence is sure to cause
severe inflammatory action and destruction of tissue, if permitted to come into contact with the circulation through an abrasion.

As lachrymal abscess is usually an affection of
adults, the aged are more liable to suffer infection
from it, through corneal abrasions, than children.

In severe cases of purulent ophthalmia, where
there is much chemosis, the lymphatic circulation
of the cornea is interfered with to such a great extent
that the nutrition is entirely repressed. Then corneal ulcer is a natural consequence.

In cases of this kind, loss of vison is almost
certain, considering the feeble nutrition of the cor-

nea, together with the action of the pyogenous micro-organisms contained in the purulent secretions which are now brought into contact with the abraded cornea.

This condition applies to all ages, and is probably the chief cause of loss of vision in ophthalmias of the purulent type, and especially in ophthalmia neonatorum.

Secondary Ulcer.—Secondary ulcer of the cornea has its starting point, without regard to cause, in some other structure of the eye, and then passes over to the cornea. Thus a secondary ulcer may be of systemic origin, as when a phlyctenular ulcer of the conjunctiva passes over to the corneal surface.

The most common varieties of secondary ulcer are the *herpetic* and the *serpiginous*.

Herpetic ulcer is a *vesicular* eruption, which first appears upon the conjunctiva, usually very near to the edge of the cornea.

Herpetic ulcers usually appear in groups, accompanied by much lachrymation, frequently, but not always, with photophobia, but without much conjunctival irritation, except at the point of contact, and with no swelling of the lids.

These groups of ulcers usually run in different directions across the cornea, often limbed shaped, as the branches of a tree, but occasionally in a straight line, either obliquely, horizontally, or sometimes in a vertical direction. As fast as the epithelium is cast off from one, another ulcer or group of ulcers makes its apperance, in the order indicated.

The vesicular eruptions in herpetic ulcers are characteristic, inasmuch as the fluid contained in them is clear, in contradistinction to the grayish matter in phlyctenular corneitis.

The covering of the vesicle soon gives way, and is replaced by an irregular but clear-cut depression in the clear cornea. This is also in contradistinction to the gray ragged edges found in phlyctenular ulcer.

Another characteristic of vesicular ulcer is that it is usually an affection of puberty, while the phlyctenular ulcer is a disease usually of childhood.

Herpetic ulcer may be primary, and have its origin occasionally upon the cornea, but this is rare. When it has its origin upon the cornea, the inflammatory symptoms are very much increased. There is pain, much lachrymation, photophobia is always present, the conjunctiva congested and the lids swollen, accompanied often by severe blepharospasm.

The herpetic ulcer, more plainly than others, is the result of an inflammatory disturbance within the territory of the fifth nerve, for the herpes zoster frontalis is usually the precedent of this type of corneal ulcer.

A careful examination of the mucous membrane of the nostrils in cases of herpetic ulcer will disclose the fact that it is also affected, being swollen, having much secretion, with irritating surfaces denuded of the epithelium and frequently covered with scabs.

Treatment.—Treatment in the early stage of herpetic ulcer, before the vesicle bursts, does not avail much, and consists mainly in the use of cocaine to mitigate the pain.

One important point in the treatment of corneal ulcer, whatever its character, is to paralyze the accommodation of the affected eye, not so much for its mydriatic affect, as to relive all strain. This is an important matter and should not be neglected.

To accomplish this purpose, a few drops of a 1% solution of atropine should be instilled into the eye once a day, unless there is a condition present indicating extreme mydriasis.

After the vesicle has ruptured, treatment should be immediately applied to the abraded surface. This consists in applying, from a small cotton holder, a 1 to 1000 solution of the bi-chloride of mercury. This application should be made once a day.

If, as it usually happens, the nostrils are implicated, the solution should be applied to the affected parts.

There can be no substantial improvement as long as the eye is painful, and it must be relieved. The best and most lasting application I have found for this purpose is the following:

R.—acid carbolic (pure), gtts, ii
      morphia sulphas,
      cocaine murias, aa, grs, iv
      glycerine, drs, ii
      aqua hamamelis, drs, vi.
      Mix and filter.
      A few drops into the eye when painful.

If there should be severe inflammatory action, hot water applications should be employed in connection with the treatment indicated heretofore.

The *serpigenous ulcer* is a creeping ulcer of the cornea, having a tendency to spread in a circular direction. It ploughs a narrow groove, usually, entirely around the circumference of the cornea, and thus disconnects its center from nutritive supply.

The nutrition thus being cut off in the manner above described, the central portion of the cornea becomes opaque and frequently sloughs, and the vision is entirely lost, or reduced to the perception of light only.

In some cases it attacks other parts of the cornea, but whatever part is affected, there is the tendency to the circular formation, these rings being sometimes small, and at other times will encircle almost the entire cornea as above described.

Serpiginous ulcer is seldom opaque, and is not usually surrounded by an opaque border, and is thus easily overlooked.

During the stage of sloughing there is no vascularization of the other parts of the cornea, but the appearance of new blood vessels, passing toward the ulcer from the conjunctiva, is the first indication of commencing repair.

Strange to relate, these ulcers, however deep, are not usually accompanied with much pain, photophobia, or lachrymation, the usual attendants of all inflammatory conditions of the cornea. This indicates that the ulcer is due to changes in nutrition caused by nerve lesion.

Occasionally, however, there are exceptional cases where the pain is intense, and the photophobia, and its usual accompaniment, blepharospasm, is very severe.

Serpiginous ulcer usually occurs in feeble, elderly people, and creeps slowly but persistently to a completion of a circle, requiring, often, three or four weeks in making the round.

Treatment.—As in herpetic ulcer, the most logical treatment in this form, is the application to the abraded surface of the bi-chloride of mercury solution, 1-1000, once a day.

The accommodation must be paralyzed, and pain must be subdued by the cocaine solution, or better, the solution of acid carbolic, morphia and cocaine, given under the treatment of herpetic ulcer.

Much has been said in the praise of the galvano-cautery in the treatment of corneal ulcers, especially of this type, but I have not found it as effective in my hands as the treatment heretofore indicated.

I have found, in this form of ulcer, that in old persons whose nutrition is impaired, the internal administration of a stimulant, as brandy, in liberal doses, is an important adjunct to the treatment.

Systemic Ulcer.—This form of ulcer is due to some affection of the system at large, the most important, on account of its frequency of which, is the *phlyctenular.*

Phlyctenular keratitis consists of the formation of a small vesicle or vesicles on the cornea. These vesicles consist of a raised portion of the epithelium of the cornea, having the appearance of a raised blister, underneath which is a serous fluid.

The vesicle is always accompanied with more or less inflammatory action owing to its situation, and the number of vesicles, as often several are found in close proximity.

The inflammatory action is also greater when the vesicles are situated near the sclero-corneal border.

In two or three days after the vesicle forms it bursts and its contents escape, and a small ulcer is formed, which is known as a *phlyctenular ulcer.*

Phlyctenular ulcers are more liable to form at the periphery of the cornea, than near its centre, although they may form on any part of the cornea, and also upon the conjunctiva.

Those single ulcers which form at the periphery of the cornea are productive of great inflammatory action, and are often dangerous, as they are liable to perforate.

Moreover, these peripheral ulcers may assume a serpigenous character, and creep along the surface of the cornea. This form is known as the *vascular* or *herpetic ulcer.*

Phlyctenular keratitis is a disease particularly of childhood, although it may, and often does, affect adults.

Whether in children or adults, these ulcers usually occur in persons of scrofulous habit and delicate constitution.

Children who have inherited syphilis are particularly liable to the phlyctenular ulcer.

The pain and inflammation accompanying phlyctenular keratitis is variable; but there is a condition known as *photophobia*, an intolerance of light, which in some cases is so severe that the patient cannot stand the least light, and he seeks the darkest place and buries his head in the pillow or bed-clothes to prevent its admittance.

This condition is not confined to children alone, but is frequent in adults also.

With the photophobia, there is also another condition, known as *blepharospasm*, or a spasmodic contraction of the lids. It is a usual accompaniment of photophobia.

Photophobia and its concomitant, blepharospasm, are due to conjunctival irritation, superinduced, usually, by corneal ulcer.

An important matter is that the amount of conjunctival irritation does not regulate the intolerance to light, for it is frequently very severe from the slightest irritation, and vice versa.

A phlyctenular ulcer in a child of strumous constitution is almost sure to bring about more or less photophobia.

The cramp or spasm which causes the blepharospasm starts in a reflex manner from the sensory facial nerves. This is easily demonstrated, as pressure upon the affected nerve often relives the spasm at once.

Treatment.—The treatment of phlyctenular keratitis in children is difficult because they are usually unmanageable, and it is often impossible to see the eye without physical force, or the use of an anaesthetic.

I prefer to give an anaesthetic when there is much resistance, especially for the first examination of the eye, for the reason, that should there be a deep ulcer, great resistance might cause perforation, which could otherwise be prevented.

In the case of an adult, a few drops of a four per cent solution of cocaine in the eye will render it capable of being very easily examined. The ulcer should be thoroughly cleansed (after the eye has been cocainized) with a small pledget of absorbent cotton placed upon the point of a small cotton holder. The cotton should be moistened in a 10-grain solution of boracic acid.

After the ulcer has been thoroughly cleansed in the manner described, its surface should be touched with a stimulating antiseptic solution. I have used the following with great satisfaction:

R, chloride zinc, grs i,
    cocaine murias,
    morphia sulphas, aa grs iv,
    acid carbolic, gtts v,
    glycerine, drs ii,
    aqua rosa, drs vi,
    Mix and filter.

Apply with a small cotton applicator to the walls of the ulcer once a day.

After the ulcer commences healing, it will not be necessary to use it so often, but it should be very carefully applied once a day until it is pretty well healed.

This application is very soothing, especially where the ulcer is painful.

In corneal ulcer, no matter what its character or where situated, the accommodation should be paralyzed with atropine. The reason for this is, that the eye should be relieved from all strain.

It has been suggested that the local effect of atropine upon the ulcer is beneficial, but I am sure that this is not the case, but the benefit comes from the loss of accommodation, and thus having the eye in a state of rest.

There is a rule however to be observed with regard to deep ulcers:

When a deep ulcer is situated at or near the center of the cornea, and we are fearful of perforation, the pupil should be widely dilated, and its edges kept as far away from the ulcer as possible, in order to prevent an intrusion of the iris into the wound, should the cornea rupture.

On the other hand, should the ulcer be situated near the periphery of the cornea, the pupil should be contracted with eserine for the same reason.

The patient should always have a soothing lotion at hand to relieve the suffering should the eye become painful. An eye can not improve as long as it is painful.

To allay the pain I use something like the following, which can be varied to suit individual cases:

R, hydrastin, gr, ss
    acid carbolic (pure), gtts, ii,
    cocaine murias, grs, viii,
    glycerine, drs, ii,
    aqua hamamelis (dist'd), drs, vi,
    Mix and filter.

A few drops should be instilled into the eye whenever it becomes painful, and continued, as often as every five minutes, until the pain ceases.

The combined anaesthetic effects of the cocaine and carbolic acid in corneal ulcer, of whatever character, is very advantageous because of its lasting effects.

The general treatment in phlyctenular keratitis should be conducted in accordance to the cause.

Invigorating measures, such as exercise in the open air, nutritious food, salt water baths as warm as can be borne, and such other remedies as are best adapted to give tonicity should be resorted to. The syrup of the iodide of iron, cod liver oil, sulphate of quinine, and iodide of potassium are indispensables.

It is important that the nasal cavities should be examined and any catarrhal condition of the nose or throat should be treated.

Photophobia is one of the most annoying complications of phlyctenular ulcer, and may continue long after all inflammatory actions have subsided.

I have often relieved photophobia in adults by the hypodermic injection of morphia over the supraorbital nerve. I would not however advise its use in children.

In young persons, in severe cases, the speediest and most permanent relief can be attained by *canthotomy*.

How does canthotomy relieve blepharospasm?

The conjunctiva of the lids is richly supplied with sensory filaments from the fifth pair of nerves. Irritation of the peripheral extremities of these nerves, by a reflex action causes cramp; pressure increases irritation; canthotomy diminishes the pressure, and therefore relieves the cramp.

I have frequently known canthotomy relieve this distressing condition in two or three days, that had resisted local treatment for months.

SUPPURATIVE KERATITIS.

Suppurative keratitis is *diffused* or *circumscribed.* In the *diffused* form the cornea loses its brilliancy, becomes steamy, assumes a grayish-white appearance, followed rapidly by a yellowish tint, which indicates the formation of pus between the lamellae.

The epithelium now disappears, and the lamellae become separated and detached by the formation of pus.

When there is an extensive formation of pus and loss of corneal substance, the intra-ocular pressure is so great that the parts are often unable to resist it, are pushed forward, and cause considerable bulging of the cornea and often perforation.

Suppurative keratitis, whether diffused or circumscribed, is always an acute affection, following very rapidly what was often thought to be simply a slight serous conjunctivitis.

Although suppurative keratitis may be the result of a corneal traumatism, such as a blow or other injury upon the cornea, or of an operation upon the eye involving the cornea, as for cataract; sometimes however, a suppurative keratitis is due to a purulent conjunctivitis, and is caused by the corneal nutrition being interfered with from the conjunctival swelling and accompanying chemosis.

Suppurative keratitis may be precipitated without any known injury to the eye, or any known cause whatever, but when it occurs spontaneously, it is usually in persons of scrofulous habit, or in old people whose nutrition is not good.

Circumscribed keratitis is first indicated by an

opaque whitish spot upon some part of the cornea. The central part of the spot soon changes its color to a yellowish tint which indicates the existence of pus.

If the pus formed is near the surface, the outer layers of the cornea break down and the pus is discharged externally, and an ulcer is thus formed.

Should the pus be formed deep in the cornea, it breaks through the layers into the anterior chamber in the aqueous humor. In this case the pus being heavier than the aqueous humor, it settles to the bottom, and forms what is known as *hypopyon*.

The layers of the cornea between which the pus is situated are frequently separated to such an extent that the pus gravitates toward the bottom of the cavity and gives it the appearance of an hypopyon. This condition is known as *onyx*, and is usually easily distinguished from hypopyon.

In hypopyon the upper level of the fluid is sharply defined, and is in a horizontal line, especially if the patient has been in an erect position for a short time. It shifts its position if the patient inclines the head from one side to the other.

On the other hand in onyx the upper portion is usually irregular. It is sometimes necessary, however, to resort to focal illumination in order to establish the diagnosis.

Treatment.—In the treatment of suppurative keratitis, the first matter of importance is to dilate the pupil. This is essential for two reasons: the first being, that the eye should be kept in a state of rest, in all inflammatory actions, if possible; then severe corneal inflammations always complicate the iris, and it is well to have the pupil so well dilated that there can be no adhesions of the iris to the lens capsule.

I am sure that a well dilated pupil places the iris in the very best possible condition to resist inflammation.

There can be very little exudation of serum or lymph from an iris whose pupil is extensively dilated, hence occlusion of the pupil, and iritic adhesions, the usual results of severe inflammatory conditions of the eye, are reduced to the minimum.

In the incipiency and during the active inflammatory stage of suppurative keratitis, warm water fomentations, as hot as can be borne, should be applied to the eye, and kept up as long as an hour at a time, two or three times a day. This process will often establish and maintain the circulation to such an extent as to cause resorption and prevent suppuration, if resorted to at an early stage.

After the fomenting is completed the eye should be covered with a slight compress of dry, heated cotton-wool. A light bandage may be used to keep it in place.

Should the abscess break into the anterior chamber, and an hypopyon be formed, if extensive, the operation of *paracentesis* should be made, by making an incision in the lower part of the cornea, near its periphery, and letting out the pus. On the other hand, if there is not much pus, and it is thin, it will absorb without doing any injury.

If the abscess forms near the surface, it should be punctured, its contents emptied, the walls of the sac treated antiseptically as in corneal ulcer from any cause.

If the general health of the patient is at fault, it should be corrected, as the conditions demand.

GENERAL CONSIDERATIONS OF CORNEAL ULCERS.

In practice we often meet with cases of corneal ulcer where it is impossible to draw a line of demarkation between the different forms already considered, but this cannot, from a practical point of view, be considered of very great importance, since the treatment resolves itself into a very few well established principles.

There is no affection of the eye more tedious to treat than corneal ulcer, no matter what the cause, nor however uncomplicated it may be, for it requires so great a length of time to complete its course.

An ulcer may be superficial or deep; the pain may be very slight or the patient may suffer severely; there may be much inflammatory action, or the inflammation may not be very marked; there may be severe photophobia and blephorospasm, or not; the ulcer may be very large, covering a great portion of the cornea, or it may be very small, scarcely notice able; it may be central, or it may occupy a point near the periphery. Such are the varied conditions of corneal ulcer.

Superficial ulcers are usually more painful than the deep, hence severity of pain, in corneal ulcer, is not indicative of great danger to vision.

Another thing in regard to corneal ulcer is, that those situated near the sclero-corneal border are usually much more painful than those which are situated near the center.

All ulcers that are not of traumatic origin, are to be regarded as systemic, and the general health must be regarded as a matter of great significance in this affection.

A corneal ulcer, however good the recovery may be, always leaves the cornea more or less impaired, no matter whether it has had its seat in the center of the cornea, or at its periphery, from the simple fact that it leaves more or less of an opacity.

Another matter with regard to corneal ulcer is this, there always remains therafter the tendency for a recurrence. This is particularly true with regard to phlyctenular ulcer, and I have observed it very often in the other forms also.

The most destructive ulcer to the vision is the *perforating*, whether it is central or not, for when the ulcer perforates, the iris or the lens, or both, become complicated.

Even if the ulcer is deep, and there is no perforation, the contour of the cornea is changed, which, together with the resulting opacity, causes very defective, if not complete loss of vision.

There is another and a greater danger from a perforating ulcer; the iris is liable to become dragged into the wound, and to become permanently adhered to the cornea, in which event, by its teasing and pulling upon the ciliary body, in its efforts to dilate and contract, it may be the cause of generating a sympathetic ophthalmia.

Should the corneal wound be large, a considerable portion of the iris, as well as the crystalline lens, may become prolapsed into the opening. In this case, as soon as the inflammatory action has somewhat subsided, the eye should be removed, for ciliary iritation is liable to occur and lead to sympathetic ophthalmia, and the loss of the other eye.

No operation for the removal or the reduction of the staphyloma should be entertained when the vision is irreparably lost, because the danger of exciting a sympathetic inflammation is so great that a very little time lost in the effort to retain the injured eye may be, and often is, fatal to its fellow.

One thing we should always observe, and it will bear repetition: the accommodation should be paralyzed immediately upon the discovery of a corneal ulcer. This relieves some strain, and thereby prevents a certain amount of inflammatory action. This should be observed in most all inflammations of the eye, except where there are evidences of glaucoma, but in corneal ulcer, and in irititis it is particularly indicated.

The position of the ulcer, especially if it is deep, is another matter of very great importance, especially should there be danger of its walls giving way. If it is central the pupil should be dilated, and the pupillary border of the iris kept as far from the wound as possible. If on the other hand the ulcer is near the periphery of the cornea, the pupil should be contracted for the same reason. This can be accomplished, with a solution of eserine, even while the accommodation is relaxed under the influence of a mydriatic.

If there is severe inflammatory action in connection with the ulcer, the eye should be bathed with hot water, as warm as can be possibly borne, for as much as an hour at a time. This should be done three or four times a day.

As the iris is liable to take on inflammatory action in case of corneal ulcer, especially if the latter is situated near the periphery, we must not allow an exudation of lymph to bind the iris to the capsule of the lens, or to fill the pupillary space, if it can be prevented. In this case the iris must be kept as widely dilated as possible, even should the ulcer be deep, for a perforation of the cornea at its border is not nearly so liable to occur as when it is more central, and does not subject the eye to so much danger as the iritis.

One of the most annoying complications in corneal ulcer is the photophobia and blepharospasm, but we must remember that this bears no relation to the extent of the ulcer, for the most obstinate cases have been the result of a very slight affection of the cornea.

The treatment is simple and usually very effective: a canthotomy, and the application of simple remedies to relieve any existing conjunctival irritation.

SECTION II.

STAPHYLOMA.

Staphyloma may be defined as a protrusion of the cornea or sclera, due to inflammatory action.

Corneal staphyloma consists of a prolapse of the iris through a wound of the cornea, generally the result of corneal ulcer.

Scleral staphyloma consists of a thinning of the scleral tissue, which gives way before the intraocular pressure.

There are two forms of scleral staphyloma, viz.: *anterior scleral staphyloma* and *posterior staphyloma*.

Anterior scleral staphyloma occurs over the region of the ciliary body. This condition is generally due to syphilis.

Posterior staphyloma of the sclerotic occurs in the posterior segment of the sclera. This condition is usual in myopia.

The *treatment* of corneal staphyloma, especially if there is much protrusion of the cornea and iris, is enucleation.

It is now generally conceded that those operations which consist in the removal of a whole or part of the corneal tissue, together with the protruding iris, and drawing the parts together with sutures, are not practical, but are often the cause of sympathetic inflammation.

When an eye is so badly injured from a staphyloma or any other cause that its vision is irreparably destroyed, especially if the iris and ciliary body are involved, enucleation is not only necessary, but urgent.

The same may be said of anterior scleral staphyloma, because the ciliary body, which is always more or less involved, is the starting point for all cases of sympathetic ophthalmia.

I am very sure that exceptions will often be taken to the removal of an eye, although irreparably blind, especially if the staphyloma be small, for by a system of puncturing the anterior chamber to allow the aqueous humor to escape, and bandaging the eye with a firm compress, the staphyloma may become reduced so much as to preserve the normal contour of the cornea; but the elements which excite sympathetic ophthalmia, that is, the adhesions of the iris in the corneal wound, still exist.

There is a form of staphyloma known as *conical cornea*, or transparent anterior staphyloma, which consists of a bulging forwards of the central portion of the cornea.

This condition should always be looked for in

near-sighted individuals, for it is often mistaken for myopia.

Conical cornea generally appears at from twelve to twenty years of age. It progresses steadily for three or four years, after which it usually remains stationary.

The cause of conical cornea is not fully established. Whether it is due to a latent inflammatory condition of the cornea, causing softening near its center, or to an intra-ocular tension, producing a yielding of the cornea, at this its thinnest portion, is not fully decided.

Rest is the only treatment, and keeping the eye fully corrected with regard to its refraction.

The various operations of trephining and excising the cornea, have not been generally accepted. Such operations are very delicate and should be resorted to with great caution.

WOUNDS OF THE CORNEA.

Wounds of the cornea, whether superficial or penetrating, are of great importance. However superficial a wound may be, if nothing more than a simple abrasion, it is liable to set up severe inflammatory action, with suppuration of the cornea and loss of vision. This is not so much from the extent of the injury, as from the introduction of septic matter into the wound; for this reason the greatest care should be exercised in dressing the wounds,

thorough asepsis should be practiced, and such antiseptics employed as are indicated.

In penetrating wounds of the cornea, the wound
should be aseptically dressed, and the iris, if protruding and not wounded, should be returned, but if
mutilated should be excised. Great care must be
taken that no part of the iris be allowed to remain
within the lips of the wound.

Atropine solution should be instilled into the
eye, and a light compress secured by a bandage.

If the penetrating wound is deep and involves
the deeper structures of the eye, as the lens, ciliary
body, etc., treatment will be more complicated.
When the lens is wounded, it frequently swells to
such an extent as to cause suppurative inflammation, and consequent loss of the eye.

FISTULA OF THE CORNEA.

Fistula of the cornea sometimes follows wounds
of the cornea either from an injury or an operation
upon the eye involving the cornea, or from a penetrating corneal ulcer, in which case there is a constant drainage of the aqueous humor. This condition
keeps the eye continually irritated, and it cannot be
used to any advantage while in this state.

Treatment consists in keeping the pupil contracted if the fistula is near the edge of the cornea, and
dilating it, if it is near the center, for the purpose of
keeping the pupillary border as far from the fistula

as possible, so as to avoid its intrusion into the opening. At the same time, the edges of the fistula should be freshened with fine forceps in order to excite the process of healing, after which the compress bandage should be applied over the eye.

ARCUS SENILIS.

Arcus senilis, known also as gerontoxon, is a hyaline or fatty degeneration of the corneal cells, appearing a little inside the margin of the cornea, and usually extending entirely around it, although more prominently marked at the upper and lower borders.

As a general thing arcus senilis is an affection of age, but it is occasionally met with in youth.

This affection is not of serious import, and no functional changes are caused by it, the cornea not losing any of its vitality because of its presence. Wounds on the arcus senilis heal as readily as on the clear cornea.

There is no treatment for arcus senilis.

CHAPTER VII.

INJURIES AND DISEASES OF THE SCLERA.

WOUNDS of the sclera may vary from a minute abrasion to a perforation or rupture.

In perforating wounds of the sclera, there is great danger of injury to the internal coats of the eye together with septic infection.

There may also be a loss of the vitreous humor, which if great, renders the prognosis very unfavorable.

Small abrasions usually heal rapidly, but in deeper wounds where the choroid and retina are implicated, the prospects of rapid healing and unimpaired vision are unfavorable.

Treatment.—Small abrasions usually heal without treatment.

When the sclera has been injured so as to make an open wound, the eye should be cleansed with an antiseptic solution. If any of the choroid protrudes from the wound, it should be removed with the scissors and the edges brought together with fine sutures.

If the wound has been large, the conjunctiva should always be sutured after bringing the edges of the sclera together.

A suitable dressing should be placed on the eye, and when soiled fresh ones reapplied.

If septic material has been introduced by the

perforating instrument, there will probably be a formation of pus and the eye entirely destroyed. In this event, timely enucleation should be resorted to, in order that the other eye may not, through sympathy, become involved.

SCLERITIS.

Scleritis is an inflammation of the sclera.

Episcleritis is an inflammation of the superficial layers of the sclera.

There is hyperaemia of a portion of the sclera and the vessels of the conjunctiva immediately above the inflamed part of the sclera become injected. The inflamed portion has a reddish-blue hue.

The pain accompanying scleritis varies according to the severity of the affection.

The duration of scleritis is usually from ten days to a month, although some cases become chronic.

Scleritis is most frequently seen in persons affected with gout, syphilis, rheumatism or scrofula.

Treatment.—The treatment consists in hot water fomentations, and where found in conjunction with syphilis, rheumatism, gout or scrofula, general remedies for such diseases exhibited.

The eyes should be protected from the cold and the patient warned not to strain.

Astringents are irritating and should not be used.

CHAPTER VIII.

SECTION I.

DISEASES OF THE IRIS.

IRITIS.

THE term *Iritis* signifies, as its name imports, an inflammation of the iris.

For convenience of study, iritis may be divided into the following forms:

$$\text{Iritis} \begin{cases} \text{Serous.} \\ \text{Plastic.} \\ \text{Parenchymatous} \begin{cases} \text{Non-suppurative.} \\ \text{Suppurative or} \\ \text{Purulent.} \end{cases} \end{cases}$$

In *serous iritis* there is an exudation of serum from the blood contained in the iris and ciliary body into the chambers of the eye. This exudation comes mostly from the posterior surface of the iris and the the anterior portion of the ciliary body.

In serous iritis the anterior chamber is often very noticeably deepened. This is owing to an excessive secretion of aqueous humor and serum, and also to an obstructed outflow of the aqueous humor into the canal of Schlemm, consequent to the swelling of the fibres of the ligamentum pectinatum, which guard the entrance to that cavity.

The increase in tension in serous iritis is due to

the above conditions; that is, to the increased inflow of lymph and serum and their obstructed outflow.

In serous iritis the pupil is sluggish in its action and somewhat dilated.

In serous iritis the exudation is mostly serum, and not so much lymph. Therefore the tendency to the formation of plastic exudations and adhesions is not so great as in other forms of iritis.

Opaque dots frequently form, in serous iritis, upon Descemet's membrane. This condition is known as *keratitis punctata*, and is caused by small coagulable particles which are formed in the lymph exudation, and which adhere to the inflamed endothelium of the cornea. These deposits settle at the lower quadrant of the posterior surface of the cornea, in a prismatic shape, the base of which corresponds to the lower margin of the cornea, the apex being directed upward toward the pupil. These deposits are frequently overlooked, because they are so excessively minute. We should, therefore, in every suspected case of serous iritis, look for them with a strong magnifying glass.

The cornea in serous iritis becomes cloudy and loses its brilliancy.

The aqueous humor in serous iritis is always somewhat cloudy, although not so much so, as in the plastic form. This is on account of the coagulable deposits which are in the exudation. As there is not so much lymph deposited in this form as in the plastic, therefore there is not so much coagulable deposit.

The inflammatory action in serous iritis is not

active, but is subacute or chronic, and of a marked recurrent type.

At first the attacks are but slight and of short duration, lasting from two weeks to a month, and then gradually subsiding; then in the course of time another attack comes on, generally of increased severity.

The increased tension of the globe becomes a more prominent feature in each successive attack.

In some cases of serous iritis profuse hemorrhages into the aqueous chambers occur from the bursting of the distended blood vessels of the iris.

Serous iritis is always accompanied by some inflammatory action, but it is not so pronounced as in the plastic form. We have the ciliary injection, pain, photophobia and lachrymation. In some rare cases, most of the above symptoms are absent.

The diminished vision in serous iritis is due to turbidity of the aqueous humor or the exudation in the pupil, or in the cavity of the vitreous.

In serous iritis the ciliary body, especially the anterior portion, is complicated. There is no direct proof of this in slight affections of the ciliary body, because it can not be seen directly, but symptoms of positive evidence of its involvement to any great extent are always present, viz:

1st. There is always tenderness on the slightest pressure in the ciliary region.

2nd. Where the inflammatory symptoms are quite marked, there is swelling or œdemia of the upper lid.

Serous iritis may, at any time, take on the plastic form, or go on to suppuration.

Although rheumatism and syphilis have been
regarded as the usual causes of serous iritis, I feel
that most cases can be referred to some ocular defect
which causes strain upon the ciliary body, such as
the errors of refraction. Outside of its being the
result of an injury, directly or indirectly (sympathetic ophthalmia), I am confident that this is often
the case.

The treatment in serous iritis is atropine. The
pupil, if possible, must be kept dilated.

Even in serous iritis, where there is usually a
small exudation of lymph, the iris may be bound
down to the lens capsule, so that the pupil will not
be influenced until several applications of the atropine have been made. In such cases a few drops
of a one per cent solution of atropine should be instilled into the eye every hour until the pupil dilates,
or in the event it does not dilate, until the inflammatory action has subsided.

PLASTIC IRITIS.

Plastic iritis is the most common form of iritis
and is due mostly to rheumatism and syphilis.

There are, however, cases of plastic iritis that
can not be traced to either of the above affections,
which are often attributed to climatic changes.

In this form of iritis there is an exudation of
plastic lymph from the blood vessels of the iris into
the aqueous chambers. In plastic iritis there are
always inflammatory symptoms, often of a marked
character, such as pain, contraction of the pupil,

conjunctivitis, circum-corneal injection of the ciliary vessels with elevation of the limbus conjunctiva and more or less chemosis.

The pain accompanying plastic iritis is not so severe as in the serous and other varieties, in fact there are cases of what is termed by some authors "quiet iritis," where it is said there is no appreciable evidence of inflammation, in which plastic exudations had been thrown out and adhesions formed without the patient suffering any inconvenience, but gradual loss of vision. These cases of iritis are of syphilitic origin, and I think a close examination of the eye during the exudative process would have shown evidence of inflammatory action.

Owing to the inflammatory condition of the eye, the cornea loses its brilliancy and becomes slightly steamy. This is caused by the involvement of the epithelium in the general inflammation.

The color of the iris changes for the same reason: that is, the involvement of the endothelium covering the anterior surface of the iris. It also changes the color because of the cloudiness of the aqueous humor and cornea, and the engorgement of the blood vessels of the iris itself.

As in all cases of iritis, the color of the iris changes from a blue to a greenish hue, and from a brown to a reddish brown.

The amount of plastic exudation varies usually as to the cause of the iritis, and also the degree of general inflammation.

If the iritis is due to syphilis, the amount of lymph thrown out is larger than in the other forms.

This is an important point in the differential diagnosis of the iritis due to syphilis, and iritis due to rheumatism. Another matter, the patient does not usually suffer so much pain when the iritis is due to syphilis, as when it is due to rheumatism. In iritis, especially, if the rheumatism is of the chronic variety the pain is usually very severe.

The amount of exudation is not always in conformity with the degree of inflammation, as a very slight inflammation may be accompanied by an exuberant exudation, and vice versa.

The anterior chamber of the eye is always deep, as in the serous variety, owing to the amount of exudation thrown out: the increase being so great sometimes as to cause an increase in the intra-ocular tension.

It is often very difficult to diagnose a plastic from a serous, or what is termed a simple iritis; happily this does not complicate matters, as the treatment for all forms does not materially differ.

The most dangerous complication in plastic iritis is adhesions of the iris to the lens capsule (posterior synechia).

In some cases of iritis the quantity of plastic lymph thrown out is enormous, and not only a part of the pupilliary border of the iris may become adhered to the lens capsule, but the whole border may become attached (seclusion of the pupil) and the pupillary space may be completely filled with the exudate (occlusion of the pupil).

The first essential in the treatment of iritis,

from whatever cause, is dilatation of the pupil, and especially is this necessary in the plastic forms.

Atropine is indicated at the earliest possible moment, and its use should be continued until all of the adhesions are broken.

If there are extensive attachments of the iris to the lens capsule a few drops of a one per cent solution of atropine should be instilled into the eye as often as once an hour, as long as the adhesions last, or until the inflammation subsides.

If the adhesions are not so extensive, the application may not be made so often, two or three applications a day being sufficient in many cases. With this treatment fomentations of water as warm as it can be borne, especially if there is much inflammatory action present.

My experience has been that there is no advantage in a stronger solution of atropine than the one per cent, and that its maximum effects can be as promptly and as thoroughly obtained by its frequent application, as by that of a stronger solution. It can be used as often as every fifteen minutes for three or four hours each day in order to detach recent extensive adhesions.

Old adhesions of any great extent cannot be broken by the use of any mydriatic, and it is in recent cases only that we can hope to effect any relief from the use of atropine.

In plastic iritis internal medication is always essential.

If it has been caused by syphilis, then the iodide

of potassium or mercury, preferably, the iodide of potassium.

I give the iodide of potassium in the form of a saturated solution, and advise the patient to take it in milk. From fifteen drops to one teaspoonful of the saturated solution in three or four tablespoonfuls of milk before meals for the adult. I have never seen this treatment disagree with the most irritable stomach, but on the other hand I have known many irritable stomachs benefitted by its use.

When we know the iritis to be of syphilitic origin, in addition to the iodide of potassium, inunctions of mercury, in the form of the mercurial ointment, rubbed in the arm pits once a day until its effects are demonstrated by a cessation of the inflammatory action, or by its constitutional effect upon the mouth or the gums.

PARENCHYMATOUS IRITIS.

Parenchymatous iritis is presented in two forms, the *non-suppurative* and the *suppurative*.

In parenchymatous iritis there is the formation of well defined masses upon the iris. These masses are termed nodules or condylomæ.

These nodules vary in size from the smallest pinhead to formations almost filling the entire anterior chamber and encroaching on the cornea.

The color of the iris changes materially at the affected part and becomes reddish-yellow or yellowish-green.

In the *non-suppurative* variety the inflammation

terminates by resolution, in which case the nodule
gradually disappears without abscess or injury of
the tissue involved. This does not often occur, as
cicatrices form in the iris substance, causing irregularities in the shape of the pupil, or displacing it.
Extensive adhesions are also liable to form between
the iris and lens capsule.

When the nodule is extensive I have known, in
a few instances, of adhesions between the iris and
posterior surface of the cornea.

In the *suppurative* form the pus gravitates to
the bottom of the anterior chamber and forms an
hypopyon.

Other portions of the eye frequently contribute
to the formation of pus at the same time the iris
is involved, especially the posterior elastic lamina
of the cornea. As this membrane is a continuation
of that which covers the anterior surface of the iris,
it is readily seen how easily it may in like manner
become implicated.

Besides affecting the posterior lining membrane
of the cornea, the suppurative form occasionally extends to the surrounding tissues of the cornea, the
ciliary body, the choroid and the vitreous.

These nodules are usually situated near the pupillary border of the iris, occasionally at its periphery, but rarely in the body of the iris.

The cause of parenchymatous iritis is mostly
syphilis.

SECTION II.

GENERAL CONSIDERATIONS OF IRITIS.

Inflammation of the iris *per se* is rare, for the reason that the uveal tract (iris, ciliary body and choroid) are supplied by the same blood vessels, and form one continuous whole. It is thus readily comprehended how all these parts may be affected at the same time. Especially is it the case with the iris and ciliary body, for either one is more liable to inflammatory action than the choroid.

The first and only symptom usual in the beginning of an iritis, from whatever cause, is a conjunctival irritation, wth its concomitant, a serous, or watery dscharge from the eye.

It is no great wonder that many physicians mistake this for a simple conjunctivitis, and prescribe some simple remedies as mild astringents, which within themselves are harmless. But the time lost is of serious moment, as frequently during this seemingly slight inflammatory condition of the eye, plastic lymph is thrown out, and the iris is often firmly adhered to the lens capsule, and the pupillary space is frequently occluded before we are aware of the real nature of the affection.

As so many eyes are lost in this manner, I desire to invite attention to a few simple points that will assist us in determining when we have an iritis.

First. When we examine an eye that is inflamed, no matter what the history of the case, we should carefully note the color of the iris, and compare it with its fellow. This is very important, especially if both eyes are not alike affected, which is not usual in iritis.

A pale blue iris becomes a dark blue or a greenish-blue in iritis, and a gray iris becomes reddish-brown.

If both eyes are inflamed, it is difficult of course to gain much information from this examination.

Second. Note the action of the pupils to light and shade.

Close both lids, and then open one of the eyes to the light. Watch the action of the pupil, and ascertain if, at the first exposure to the light the pupil was dilated, and if it at once contracted when the light entered the eye. This being the case, iritis can, as a rule, be excluded.

On the other hand if, after both eyes have been shaded, the pupil remains stationary, and upon exposure to the light it is difficult to determine whether it contracts, or we can observe no perceptible movement of the pupil to the effects of light and shade, then we must expect iritis.

It must be remembered that the pupil is not always contracted in iritis. In the serous form it is always dilated, but not to a great extent; but the inactivity of the pupil is as persistent in this form as in the plastic.

If the iritis is due to a traumatism, the cause should be removed in case it continues to exist; if a foreign body is embedded in the iris, it should be removed; if a portion of the iris is intruded in the wound, it should be liberated or removed, and if

contused, it should be excised; if the lens is dis-
dislocated, it should be removed, and if the globe is
irreparably wounded, it should be enucleated without delay.

There is no advantage in retaining a badly lacerated globe, and especially if there is any evidence
of extraneous matter within.   The retention of a
badly wounded eye is often the cause of total loss
of vision of its fellow through sympathy.

To return to the serous and plastic forms of
iritis, as a local treatment *atropine is first, last, and
always.*

As I have indicated heretofore, there is no advantage in a stronger solution of atropine, than a
1 per cent.   If the indications are urgent, then its
frequent use; if not so significant, then its application should be regulated accordingly.

Besides atropine, we have duboisin, a more powerful mydriatic, that has been used in extreme cases,
but as it is very liable to produce toxic symptoms, it
should not be resorted to, except by one who has had
some experience in its use.   For myself, I have had
such experience with duboisin, that I am content
to be satisfied with the medicinal virtues of atropine.

Whether in the serous or plastic forms of iritis,
internal treatment is of paramount importance.   If
due to rheumatism or syphilis the iodide of potassium is alike indicated.

I always use the saturated solution, and give it
in milk.

Whenever there is an exudation of lymph,
whether serous or plastic, the iodide of potassium is

admissable.   If the lymph is more of the serous
character, the dose need not be so large, but if of
the plastic form, then it should be pushed to its
utmost extent.

While morphia and other narcotics are permis-
sible in the general treatment of iritis, especially if
the pain is severe, their use is often detrimental, in
that they check the secretions and derange the digestion.

It may not be generally known, but it is a fact
nevertheless, that the use of the iodide of potassium,
as heretofore indicated, will relieve the pain and give
rest, in iritis, when narcotics will not, and without
disturbing the system at large.   It is especially indicated in the supra-orbital neuralgia, which is so
frequent an accompaniment of iritis, especially of
the serous variety.

Salicylic acid and salicylate of sodium, which
were in former years so highly extolled, especially
in that form of iritis due to rheumatism, need only
be mentioned to be condemned, as their effect in
disturbing the digestion is too well known to permit
their employment.

The old methods of cupping, leeching and blis-
tering are of no value in iritis, and the patient should
not be subjected to any of these antiquated procedures.

Poultices are valueless, and for aseptic reasons
alone should not be resorted to in any affection of
the eye.

During the active inflammatory stage of plastic
or purulent iritis, the application, over the eye, of

hot water fomentations is beneficial, as well for its
sedative effects, as to hasten absorption.

The cloths should be wrung out of warm water,
as hot as can be borne, and applied as often as every
five minutes for an hour at a time, as often as two
or three times a day.

Cocaine if applied frequently will affect the
deeper structures of the eye, and is very beneficial
in conjunction with atropine when the eye is painful.
I would recommend in such cases equal parts of a
one per cent solution of atropine and a four per
cent solution of cocaine. A few drops in the eye
every five minutes until the pain is relieved; afterwards as often as necessary to relieve the pain and
keep the pupil dilated.

Paracentisis of the anterior chamber should not
be performed, unless the chamber has an accumulation of pus.

It is a dangerous procedure to make an operation upon a badly inflamed eye, and should not be
resorted to only as a *dernier ressort.*

Iritis is more common in adults than in children, and for that reason great care must be taken
in diagnosing the affection from glaucoma, the latter
being a disease usually of advanced life.

The application of atropine in glaucoma is dangerous, and very liable to set up a destructive inflammation.

The conjunctival sac should be flushed two or
three times daily with an aseptic solution, as the bichloride, 1-5000, or the boric acid solution. This is
particularly necessary if there is a high stage of in-

flammatory action. This procedure is highly beneficial, not only in iritis, but in any inflammatory condition of the eye, as the conjunctival sac is a receptacle for all of the pus, mucus, and other inflammatory products which are thrown off, the retention of which retards the resumption of the normal functions of the eye.

Immediately before the flushing a few drops of a four per cent solution of cocaine should be instilled into the eye—then the flushing has not the least disagreeable effect upon the patient.

In iritis, from whatever cause, during the active stage, the eye should be covered with heated pledgets of dry cotton or wool, as light as possible, so that the eye may be kept warm with the least pressure possible.

CYSTS OF THE IRIS.

Cysts of the iris are very rare, and are usually the result of an injury. They usually appear as transparent vesicles on the surface of the iris, either attached by a broad base or a pedicle, usually however by the former.

The treatment is to excise that portion of the iris to which they are attached.

CONGENITAL MALFORMATIONS OF THE IRIS.

The following conditions may be regarded as the congenital malformations of the iris: aniridia, coloboma, corectopia, heterochromia, persistent pupillary membrane, and polycorpia.

Aniridia is that condition of the eye in which the

iris is wanting. This defect is rare. It is also known as *irideremia*.

Aniridia may affect both eyes. When there is entire absence of the iris, the ciliary processes can be readily seen.

Those persons who are affected with absence of the iris suffer very much from the effects of the light, for the protection of which the *stenopaic glasses* should be prescribed.

Coloboma is that condition of the iris in which there is a cleft or fissure, resembling an artificial pupil.

This fissure of the iris is usually directed downward, and is more frequent in both eyes, than in a single eye. This cleft is sometimes continued into the ciliary body and choroid.

Corectopia is an eccentric position of the pupil.

The normal pupil is a little below the center of the iris, and to the nasal side, but is not readily observed. In corectopia the eccentric position is very easily seen.

Heterochromia is that condition in which there is a difference in the color in one iris, or that condition in which the color of one iris differs from that of the other.

This condition is due to a want of uniform pigmentation, and is not significant of any pathological state. Called also, *heterophthalmus*.

Persistent pupillary membrane is that congenital condition of the iris in which the fibres are seen to arise from the anterior surface of the iris and pass across the pupil to the opposite side.

These fibres have the appearance of threads, and are either single, or often composed of many, usually in a group.

When the pupil is contracted the fibres relax and float about in the aqueous humor; when, however, the pupil is dilated, the fibres are straightened and tightly drawn across the pupil.

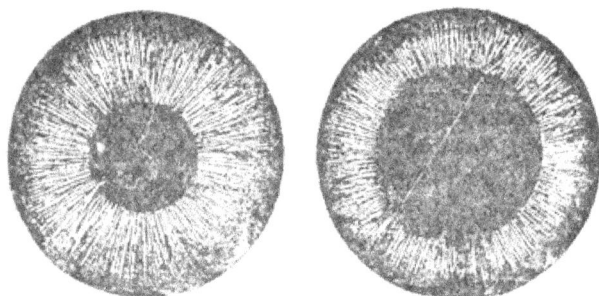

Persisting Pupillary membrane. 1. Pupil contracted. 2. Pupil dilated. Wickerkiewicz.)

FIGURE 23.

This defect is not easily detected, and as it does not usually interfere very much with the vision, the risk of an operation in removing it, is not justifiable. I have known these fibres, in one instance, to become detached, by the stretching of them, in wide dilatation of the pupil.

Persistent pupillary membrane is probably the result of an incomplete resolution of the embryological membrane, which closes the pupil in utero.

Polycorpia signifies more than one pupil, or a number of pupils in the same eye.

Where there are many pupils, it is usually

caused by the remaining fibres of persistent pupillary membrane, crossing in different directions.

IRIDECTOMY.

This operation consists in the excision of a part of the iris, and is indicated in many affections and conditions of the eye, as in glaucoma, chronic iritis, exclusion and occlusion of the pupil, in some forms of cataract as the pyramidal, and in central opacity of the cornea.

The instruments necessary for the performance of an iridectomy are a speculum, fixation forceps, a Graefe's cataract knife, a pair of iris forceps, which may be either curved or straight, and a small pair of scissors curved on the flat.

It is not necessary to use an anaesthetic, except for children or when the eye is very sensitive from injury or a long continued inflammatory condition.

If we do not give an anaesthetic, then the eye must be well cocainized. A four per cent solution of cocaine dropped into the eye every two minutes for a quarter of an hour, will so thoroughly anaesthetize it, as to render it entirely insensible to the procedure necessary in this operation.

I use the Graefe cataract knife to make the corneal incision, instead of the lance-shaped knife generally recommended in this operation.

The incison is made in the sclero-corneal tissue, about one line posterior to the edge of the clear portion of the cornea. This incison should be made in the same manner as that for the extraction of cataract, but it should be smaller. When the incision is completed, it is usual for the iris to prolapse, especially if it is not adherent, when it can be grasped with the iris forceps and gently pulled out and cut off with the scissors as close as possible to the ciliary insertion. This is very important, especially in glaucoma, in which case the best result is secured by a large iridectomy well into the ciliary body.

We should be very careful that the wound is left clear; that is, that no portion of the iris is left within its lips, and that it is perfecly cleared of all extraneous body. There is nothing that will contribute more to a long continued irritation of the eye following an iridectomy than the intrusion of a portion of the iris.

If the iris does not prolapse immediately after the incision is made, then we must introduce the forceps or the iris hook and draw it out; but great caution must be observed in this procedure not to wound the crystalline lens, as its opacity would be sure to follow.

If the iris is not adherent, the operation is very simple, and is usually easily performed, but if there is an adhesion of the iris to the lens capsule, or if plastic lymph has been thrown out and organized by the formation of a membrane in the pupillary space, then the operation is very difficult, especially when it is made for the purpose of securing an artificial pupil.

A light compress bandage should be placed on the eye and quiet maintained for several days.

IRIDOTOMY.

Iridotomy consists in making an incision into the iris where the pupil has been closed by inflammatory deposit, with the expectation that the edges will retract sufficiently to make an opening large enough to serve as a pupil.

Various instruments have been devised for this purpose, such as the sickle-pointed knife, and the scissors of M. de Wecker.

CHAPTER IX.

DISEASES OF THE CHOROID.

CHOROIDITIS.

CHOROIDITIS, an inflammation of the choroid. In all cases of choroiditis there is an exudation of some character, into the substance of the choroid, hence the term *exudative choroiditis*.

The exudate may be *serous, plastic* or *purulent*.

The serous and plastic varieties belong to what is known as the *non-suppurative* form, and the purulent to the *suppurative* form, therefore, non-suppurative and suppurative choroiditis.

The *non-suppurative* form may undergo resorption, but the choroid is always more or less impaired at the point of the exudate.

After the resorption of the exudate there always remains, at its situation, a denuded, an atrophied or a pigmented spot.

The above conditions are brought about in this manner: The exudate having become resorbed, absorption, retraction or cicitrization of the involved tissues takes place, leaving the sclerotic exposed, or the choroid atrophied, or newly-formed connective tissue at the seat of the exudation.

Should the retina become involved with the exudation, as it often does, it also undergoes the same changes in its structure as with the choroid, that is, it may become entirely absorbed or atrophied, or newly-formed connective tissue deposited at the seat of the exudate.

If the inflammation remains confined to the choroid, then there is no outward indication of inflammation observable. The only manner in which the disease manifests itself is observable to the patient, in the defect of his vision, or through the ophthalmoscope to the physician.

Choroiditis is very chronic, owing to the time it takes an exudate to resorb, together with the process preceding the full completion of atrophy and cicitrization of that part of the choroidal tissue in which the exudation occurred.

The tendency of choroiditis is to recur, and it is not infrequent for cases to terminate in partial, if not total loss of vision.

Choroiditis may, by continuity of structure, pass over to the ciliary body and the iris.

The causes of choroiditis are syphilis, scrofula, rheumatism and meningeal and cerebral lesions. It is sometimes congenital, and is almost always present in very high degrees of myopia.

For convenience of study the symptoms of non-suppurative choroiditis may be divided into the *subjective* and the *objective*.

The *subjective symptoms* of non-suppurative choroiditis are:

1.   Dimness of vision.
2.   Distortion of images (metamorphosia).

3. Sensations of sparks, bright spots, balls of fire, sparkles of light, etc. (photopsia).

4. Black spots floating in the field of vision (scotoma).

The *objective symptoms* are:

1. Yellowish spots, indistinctly outlined. (Observed only during the continuance of the exudate, and before complete resorption).

2. Light colored spots. (Observed during the process of resorption, at about the time it is completed.)

3. White spots. (Complete atrophy of pigment and resorption of exudate.)

4. White spots, either dotted or lined with pigment. (Incomplete atrophy of pigment.)

5. Floating bodies of black flakes in the vitreous. (observable only in the advanced stage of choroiditis).

Contrary to almost all other affections of the eye involving the permanent loss of vision, it is not usual for the patient to suffer pain in the non-suppurative form of choroiditis; indeed it often occurs that the patient loses an eye from this affection without being aware that it has been in any manner diseased, until he accidentally discovers that his vision in that organ is materially impaired or entirely lost.

Such a condition, however, is liable to occur only when the disease affects but one organ, but as choroiditis generally affects both eyes sooner or later it is not usual for it to affect an eye for any considerable length of time before it is discovered.

Vision in choroiditis being more or less disturbed, the patient consults his physician, not because he is suffering pain, but because of the gradually increasing dimness of sight, which is more or less annoying, owing to the extent of the exudation and the portion of the fundus covered by it.

Should the exudate be deposited in the region of the macula lutea it is known as *choroiditis centralis*, and the vision is usually disturbed by it to a very great extent.

If the exudation is distributed generally over the fundus, it is termed *choroiditis disseminata*, with more or less loss of vision, although not so much as in choroiditis centralis; however, in choroiditis disseminata it frequently occurs that the region of the macula lutea is not disturbed to a very great extent, and central vision may be fairly clear, while that of the other portions of the retina is considerably deranged.

Besides the dimness of vision, there is a sensation of glimmering. This condition is almost constant in the incipiency of exudative choroiditis, and is caused by the imperfect manner in which the retina at the point of exudation receives impressions of images.

Metamorphosia, or the distortion of images, is a frequent condition in exudative choroiditis and is caused by the retina being raised at the point of exudation, the regular contour of the fundus being thus interfered with by the exudate in the choroid.

Photopsia, or the sensation of sparkles of light and balls of fire before the eye, is a very evident

symptom of irritation of the retina. This sensation
is present as well when the lids are closed as when
the eyes are open, and continues a variable length
of time, but is more marked during the incipiency
and during the inflammatory condition of the choroid which extends over a period often of several
months.

It is impossible for the choroid to be affected to
any great extent without injury to the retina, and
especially is this the case when there is an extensive exudation into the choroid, because the surrounding portion of the retina also becomes infiltrated with this exudation, in which manner choroido-retinitis is established.

As exudative choroiditis often complicates the
retina, it frequently happens that defective vision
in some portion of the retina occurs as a result of the
choroiditis. This condition is known as *scotoma*,
and signifies that a certain portion of the retina is
insensible to light, which is indicated by the presence, to the patient, of a black speck before the eye.

The characteristic peculiarity of a scotoma is its
moving with the eye, and not floating before it, as
in *muscae volitantes;* and this distinction must be
particularly observed, for often very much depends
upon our diagnosis before making an ophthalmoscopic examination, as the defect in the retina, which
causes the scotoma, may be so obscure as not to be
detected, even with the ophthalmoscope.

In scotoma the dark spots observable to the patient, although they may vary much in form and
figure, are constant and have well defined shapes,

and if there are more than one, they preserve a fixed relation to each other, while in muscae volitantes the figures vary in shape, are not constant, and if there are several, have no fixed relation to each other, but float about promiscuously among themselves.

As the muscae volitantes following choroiditis is a symptom which is due to floating opacities in the humors of the eye, especially in the vitreous humor, it is easily detected with the ophthalmoscope. This condition is occasionally the result of a hemorrhage which frequently accompanies the choroiditis.

It should be understood that a scotoma is not always the result of a choroiditis, for a portion of the retina may be imperfect or wanting, congenitally, or it may be the result of a rent in the retina, caused from a blow, or from a hemorrhage or an inflammation of the retina.

The objective symptoms, in exudative choroiditis are demonstrable with the ophthalmoscope only, the most important of which, in the early stage of the exudative process, are the yellowish spots at the point of the exudate.

There is always one peculiarity with regard to these spots, and that is, the outlines are decidedly indistinct; in fact the normal tissue is so blended with that which contains the exudate, that it is impossible to appreciate the point of transition.

After resorption of the exudate begins and there is beginning atrophy of the choroidal tissue, the outlines of the spots are somewhat better defined, and the spots themselves become whiter, are, in fact, a yellowish white, and their edges, although irregular in outline, are easily recognized.

As the resorption of the exudate continues, atrophy of the choroidal tissue is inaugurated, and continues until the spots are changed to almost a pearly white.

In some cases atrophy of the choroidal tissue is never completed, and there will always remain depositions of unabsorbed pigmentary tissue, generally around the edges of the spots, although occasionally in or near their centers.

If the choroiditis is uncomplicated, the retinal vessels are seen passing over the atrophied spot. If they do not pass over, but run only to the margin of the patch, then the retina has become involved also.

Sometimes, in the very early stages of choroiditis, the vitreous seems to be full of dust like floating opacities; these are due to the exudation of inflammatory material, and are readily seen with the ophthalmoscope, and move among themselves when the eye is moved.

The choroid and the iris being so closely connected, forming as they do, with the ciliary body, one continuous whole, the causes of the different forms of choroiditis may be regarded with considerable reasonableness, in the same light as those of the different forms of iritis.

Like iritis, exudative choroiditis is usually associated with some depraved state of the constitution, as syphilis, rheumatism and gout; but by far the most frequent cause is syphilis. It has always appeared to me a most fortunate thing that the great majority of the cases of choroiditis is of syphilitic origin, for syphilitic choroiditis is most amenable to treatment, and a rapid improvement is usually obtained in a remarkably short time, with the proper remedies.

Treatment.—The first and most important matter in the treatment of exudative choroiditis of the non-suppurative form, is rest for the eyes and their protection from the bright light. The first can be accomplished by wearing smoked or colored glasses, and the avoidance of the light by remaining in doors as much as possible.

In regard to rest for the eyes, they should not be strained by any work requiring their use. This is very important, and should be strictly enforced, by paralyzing the accommodation if necessary.

The symptoms in the incipiency of choroiditis and glaucoma are so nearly analogous, that it is necessary to differentiate them before a mydriatic is applied to the eyes, if we do not desire to suffer the reproach of having precipitated an attack of glaucoma.

If we have demonstrated, with the ophthalmoscope, that an exudation has taken place, then we are safe in using the mydriatic. If not, then we would better depend upon the honesty of the patient in observing our instructions.

As soon as it is possible to make an examination of the eye with regard to its refraction, it should be done, for I am inclined to think that many of these cases of choroiditis, that make their appearance in persons of about forty-five years of age, are due to the strain consequent in the beginning of presbyopia.

In refractive error, all the strain, if any, is upon the ciliary body; and as the ciliary body is so closely connected with the choroid, any strain upon it will certainly compromise the choroid, and more especially if there is a latent tendency to a choroiditis.

All defects in the refraction should be fully corrected, and examinations made from time to time in order that a change in the glasses may be made when neccessary.

The general health is a matter of great importance, especially in elderly people, and remedies suitable to each particular case must be prescribed. The bowels must be well regulated, and in cases of anaemia, suitable tonics must be given.

Should we have cause to suspect syphilis, the iodide of potassium or mercury should be administered. I am very much in favor of the iodide of potassium in large doses, in the young as well as in the feeble and the aged. I usually prescribe for the adult one-half teaspoonful (30 grs.) of the saturated solution, in two or three tablespoonsful of milk, before meals. In children from two to ten years of age, I prescribe from fifteen to twenty drops of the same.

The iodide of potassium should always be given in the saturated form, and should be taken in milk. I have never known the iodide of potassium given in this manner, to disagree with the most irritable stomach; in fact I have known of many irritable and weak stomachs to have been cured by the use of this remedy in the manner prescribed.

Inunctions of mercurial ointment should be practised, especially if the patient is robust and young. The feeble and the aged cannot withstand its depressing effects, and it should not be resorted to in those cases.

Pilocarpine, for the purpose of producing diaphoresis in cases where there is much opacity of the vitreous, has been recommended, but it must be administered with very great care. For the aged there is no tonic that equals strychnia.

Suppurative choroiditis. The suppurative form of choroiditis is characterized by an exudation of pus in the choroidal tissue, or between the choroid and retina, which rapidly extends, on account of contiguity of structure, to the ciliary body and iris, as well as to the vitreous humor.

Unlike the non-suppurative form, the patient, in suppurative choroiditis, suffers severe pain. There is extreme tension, the iris is pushed forward, and the anterior chamber is very shallow.

If the eye is examined in the incipiency of the affection, before the humors become clouded, the ophthalmoscope will reveal the purulent mass pushed forward into the vitreous, giving a yellow reflection.

The vision is gradually lost, the lids become oedematous, there is chemosis of the conjunctiva, the cornea becomes opaque, and pus is deposited into the anterior chamber, forming an hypopyon.

Because of the involvement of all the tissues of the eye and its appendages, in the inflammatory action, the globe is pressed with such great force against the swollen lids, that neither the eye nor the lids can be moved.

Suppurative choroiditis is usually caused from wounds of the eye, either from accident or by an operation, especially that for the extraction of cataract. However, since antiseptics have been resorted to, and aseptic precautions have been so strictly observed in all operations upon the eye, suppurative choroiditis, from operations, has greatly diminished.

In some rare cases suppurative choroiditis is caused by an extension of inflammation from the cornea and the iris, as in sloughing ulcers of the cornea. It also occurs in different forms of septicaemia, and in children in cerebro-spinal meningitis.

Treatment in suppurative choroiditis avails little, except, perhaps, in that following cerebro-spinal meningitis, from which a few cases of recovery are upon record; and in these cases it is really a question as to whether the exudate was not plastic, instead of purulent.

Fomentations as hot as can be borne, should be applied frequently, and a free incision into the sclerotic, so as to give passage to the pent up contents, should be resorted to, at the earliest possible moment, in order to relieve an extended course of suffering.

The subsequent treatment consists in dressing the eye aseptically. The inflammation now gradually subsides, leaving the eye in that condition known as *phthisis bulbi.*

CHAPTER X.

DISEASES OF THE CILIARY BODY.

CYCLITIS.

THE ciliary body being so closely connected with the iris and the choroid, can scarcely avoid being more or less implicated, should either one or both of the latter suffer inflammatory action; neither can there be an inflammation of the ciliary body without the choroid and iris being more or less affected; therefore, inflammation of the ciliary body is seldom found to occur without a similar condition of either the iris or choroid, or both, except in case of direct injury to that structure.

In cyclitis, there is always tenderness over the "danger zone" and injection of the blood-vessels in this region.

The aqueous humor is cloudy, and flocculi of lymph, or pus, and sometimes blood are seen in the anterior chamber.

The vision is always more or less impaired owing to the extent of the turbidity, not only in the aqueous humor, but in the vitreous also. The vitreous humor as seen with the aid of the ophthalmoscope, is more or less filled with floating opacities.

Cyclitis is a dangerous affection and may result in a destructive suppuration and atrophy of the globe, under the best advised treatment.

There is another danger besides suppuration, and equally fatal to the integrity of the eye, and that is a thinning of the walls of the sclerotic over the ciliary body, consequent to the long continued inflammatory action, and the formation of a ciliary staphyloma. In case this occurs, the removal of the eye in order to prevent sympathetic inflammation of its fellow, is imperative.

Cyclitis occurring with, or as the result of choroiditis or iritis, demands the same treatment as those affections, and like them, may entirely recover.

CHAPTER XI.

SECTION I.

DISEASES OF THE RETINA.

Retinitis.

RETINITIS, an inflammation of the retina.

Retinitis may be *primary* or *secondary.*

In *primary retinitis* the affection begins in the retina.

In *secondary retinitis* the retina is implicated through an inflammatory process which begins in some other part of the eye.

Primary and *secondary,* as here used, should not be confounded with the terms *idiophatic* and *symptomatic,* for the retina may become affected primarily, that is, before other parts of the eye are affected, and the affection would not be idiopathic, but may proceed from some prior disorder of the system at large.

Idiopathic retinitis, that is a retinitis not preceded or occasioned by some other disease, is very rare, and seldom occurs. It is usually caused from exposure of the eye to a very bright light, or from severe and long continued strain upon the eye in observing a near point.

Symptomatic retinitis, or that which is the result of some constitutional disorder is usual, the most common of which are the following:

Albuminuric retinitis.

Syphilitic retinitis.

Glycosuric retinitis.

Leukemic retinitis.

Another form which is very common, pigmentary retinitis, or *retinitis pigmentosa*, has been attributed to constitutional affections, chiefly to inherited syphilis, but this view is not generally accepted.

Albuminuric retinitis is a form of retinitis which is due to albuminuria, and is the most common form of retinitis.

It frequently occurs that the first symptom of albuminuria is found in the appearances of the retina as is often revealed by the ophthalmoscope, in search of a cause for the attending defective vision.

Albuminuric retinitis is due to morbid changes in the connective tissue fibres of the retina, occasioned by a toxic effect, from changes in the blood.

Loss of transparency of the retina is one of the first ophthalmoscopic impressions observed, and is manifest in pathological changes, just as a loss of transparency in the other transparent tissues, as the cornea, the aqueous, the lens, and the vitreous; consequently, minute alterations in the retina are discovered very early in albuminuria, if a careful examination is made with the ophthalmoscope.

The connection between kidney lesion and retinitis is very obscure, but there is doubtless a development of disease, or a morbid change affected in the walls of the retinal vessels in consequence of an altered composition of the blood, which change results in inflammation itself.

Albuminuric retinitis is of serious import, and patients suffering from it do not survive long. Fuchs claims that within his experience they usually succumb within a year.

The primary effect of albuminuria upon the retina is inflammatory, after which there is a retinal degeneration, especially if the albuminuria continues for any great length of time.

There is a form of albuminuric retinitis that is very transient, which has been denominated *albuminuric amaurosis*, and which is observed in pregnancy and other conditions producing acute nephritis. It is really a question whether the albumen in the blood causes a transient congestion of the retinal vessels, or if blindness is not due to its toxic effect upon the nerve filaments.

In albuminuric retinitis both eyes are affected, but almost always one to a greater extent than the other.

The first subjective symptom of albuminuria is frequently the defective vision, and for this reason, a very careful ophthalmoscopic examination should be made in all cases of decreasing vision which are obscure, and cannot be attributed to a refractive error, or other obvious cause.

Treatment.—The local treatment in albuminuric retinitis is simply rest for the eyes, and the use of smoked glasses to protect them from the bright light. The general treatment should be directed to the kidneys.

Syphilitic retinitis is the next most common form of retinitis, if it is not the first.

Syphilitic retinitis is generally associated with choroiditis, hence *choroido-retinitis*, and is subordinate to it, having originated in the choroid; however primary syphilitic retinitis is occasionally met.

Syphilitic retinitis occurs in both the congenital and acquired forms of syphilis. It may affect one eye alone, but it generally affects the second eye sooner or later, and may occur several months or as much as a year afterward.

In cases of inherited syphilis, the affection most often occurs between three years and fifteen years of age, although it has been seen as early as six months.

In acquired syphilis the retinitis usually appears in from six months to two years after infection.

Defective vision is usually very marked from the very incipiency of the affection, and if treatment is neglected, often continues until there is total blindness.

The tendency of syphilitic retinitis is to sudden relapses and aggravation of the affection after temporary amelioration or improvement. Because of this inclination to recur, syphilitic retinitis usually continues in its course from bad to worse, until useful vision is entirely lost.

Treatment.—The treatment in syphilitic retinitis is local and general.

As in retinitis from any cause, very little can be accomplished by local means except to cover the eyes with smoked glasses in order to protect them from the bright light. Excessive strain in attempting to read or to do fine work with the eyes, or

in fact any near work should be avoided, and the
eyes given complete rest.

If seen early, syphilitic retinitis, like syphilitic
choroiditis, is very much benefited by an alterative
treatment, but it must be energetic, and because of
the tendency to relapses and exacerbations it must
be continued as long as possible. I am sure that
many of these cases can be held in suspension or
temporary extinction, but the spark is there, and can
be fanned into a flame on the least provocation, such
as a neglect of the treatment, the abuse of alcoholic
drinks, or overheating from any cause.

The administration of the iodide of potassium
internally, and inunctions of mercurial ointment,
as recommended in choroiditis, should be resorted
to. The iodide of potassium must be continued as
long as it agrees with the stomach, and the inunctions until there is evidence of ptyalism, when their
use should be discontinued until the effects pass off,
when they can again be renewed.

Glycosuric retinitis, or *diabetic retinitis*, is due
to the toxic effects of glycosuria upon the retina.
It is a very rare affection, occurring principally in
old people.

The ophthalmoscopic appearances of glycosuric
retinitis are so similar to those of the albuminuric
form that it is impossible to distinguish them by this
means alone, and recourse must be had to urinalysis.

Treatment.—As the retinitis is due to the general
health, treatment must be directed to the diabetes.
The only local means employed are those which are
indicated in the other forms of retinitis; the eyes

must be protected from bright light by smoked
glasses, and all strain must be avoided.

Leukemic retinitis is due to an altered state of
the blood, marked by an excessive and permanent
increase in the white corpuscles of the blood.

Vision may or may not be affected to a very
great extent. The tendency in this form of retinitis
is to hemorrhages, which may cause complete blindness.

Leukemic retinitis is very rare, as in only about
one-fourth of the cases of leucocythemia is the retina
in the least affected.

Treatment.—The treatment must be directed to
the general health, with the same observance in protecting the eyes from bright light and strain, as in
other forms of retinitis.

Pigmentary retinitis, or *retinitis pigmentosa*, is an
affection of early life, progressive but chronic, often
requiring many years to complete its course.

The inflammation attending this affection is not
very marked, in fact it has been asserted by some
authors to be non-inflammatory in its character, the
retinal lesion being the result of an atrophy or pigmentary degeneration.

The most prominent symptom in pigmentary
retinitis is that condition known as hemeralopia, or
night blindness, in which the patient sees much
worse at night, or in a dim light, than his vision
when in bright sunlight would seem to justify.

Pigmentary retinitis attacks both eyes simultaneously, is most frequently observed in males,
and as before mentioned is essentially a disease of

childhood, but occasionally it presents itself at from twelve to fifteen years of age.

The causes of retinitis pigmentosa are obscure, and as it often affects several members of the same family, heredity is regarded as an influential agent. It has frequently been found in families where there were prevalent defects in intellect, or several members were deaf and mute.

Treatment.—Treatment is of doubtful benefit. Electricity and hypodermic injections of strychnia have been resorted to by some, who have claimed a temporary improvement in vision, but this may be ascribed probably to the inspiring hopes of the patient in anticipation of a favorable result.

If there is a history of syphilis, the alterative treatment indicated in the other forms of retinitis should be resorted to.

So many varieties of retinitis have been recognized by authors that they are perplexing to the student, and serve to confound rather than make plain a subject which can be simplified to such a degree as to be considered under the five common and generally recognized forms already given, and which embrace fully 95 per cent of all cases of inflammation of the retina.

Beside the varieties already mentioned, there are the *purulent, hemorrhagic, apoplectic, centralis, proliferans,* etc., the consideration of which is not justified, inasmuch as they are rather conditions of common forms of retinitis than distinct affections. For instance, a hemorrhage is liable to occur in any of the common forms of retinitis; but for this reason alone, it should not be classified as a special variety.

The same may be said of the purulent, which is rather the course or the result of one of the common forms of retinitis; the centralis, which indicates the point of attack; and the proliferans, to the peculiarity of the stria formed from the connective tissue in the retina, as a result of an inflammatory action.

Another justification for simplifying the varieties of retinitis, is that the treatment in all forms is so general. It resolves itself into a very small compass: rest to the eyes, protection from bright light, attention to general health.

Ophthalmoscopic appearances as observed in the different varieties and conditions of retinitis, in the order of their usual occurrence:

ALBUMINURIC RETINITIS.

1. Hyperaemia of the papilla and of that portion of the retina immediately surrounding it, as is evidenced by a dull haze over this portion of the fundus.

2. Hemorrhages on and in the region of the disk.

3. After the disease is well established, small white spots, collected in groups around the yellow spot.

SYPHILITIC RETINITIS.

1. Slight hyperaemia of the papilla, and a considerable portion of the fundus, but most marked

about the region of the yellow spot, and around the edge of the disk.

2.   Hemorrhages occasionally, but rarely occur.

3.   "Dust like" opacities in the vitreous.   These opacities are generally diffused, often filling the entire humor, but there are present sometimes, large flake-like or membranous opacities.   Opacities in the vitreous are pathognomon of syphilis.

4.   Small white dots occasionally occur about the macula lutea, but generally very late in the affection.

GLYCOSURIC RETINITIS.

1.   Papilla very pale with margins indistinct, with slight opacity of the fundus of the retina, and along the course of the retinal vessels.

2.   Small retinal hemorrhages about the yellow spot, usually star-like in appearance.

3.   Opacities in the vitreous, usually of hemorrhagic origin.

LEUKEMIC RETINITIS.

1.   Yellowish, rounded, hemorrhagic spots in the region of the yellow spot, and at the periphery of the fundus.

2.   Small white spots and white streaks along the course of the retinal vessels, due to accumulations of leucocytes, which have passed bodily through the vessel walls.

3.   Fundus paler than normal, owing to the altered condition of the blood.

PIGMENTARY RETINITIS.

1. Star-like intercommunicating spots of pigmentation, first in the periphery, but gradually advancing in the protracted course of the disease, mostly along the course of the vessels, toward the yellow spot.

2. The papilla of grayish, yellowish, waxy appearance.

3. The arteries and veins both very small and threadlike in appearance.

4. Opacities in the vitreous, and posterior polar cataract often present in the later stages.

SECTION II.

DETACHMENT OF THE RETINA.

The retina may become detached from the choroid, either from the effects of an injury upon the eye, as a blow, or from the effects of disease of that organ.

The diseases which are liable to effect detachment of the retina, are retinitis, especially the albuminuric form, choroiditis, diseases of the vitreous, particularly those which result in an atrophy or wasting away of the vitreous, and intraocular tumors, principally sarcoma of the choroid.

A sudden loss of the vitreous, either as a result of an injury or an operation upon the eye, often causes a detachment of the retina; and it has been known to occur from the sudden loss of the aqueous humor, as in cataract operation.

Detachment of the retina may be *complete* or *partial*.

In *complete detachment* of the retina, the whole of the retina is dragged away from the choroid, except that portion around the optic disk.

In *partial detachment*, only a part of the retina is separated from the choroid, which may vary in extent from a very small portion to that covering a very large area.

In case the detachment follows a blow or an injury of the eye, the intervening space is usually filled with blood, or a bloody fluid; when it follows disease the subretinal fluid is usually clear.

The amount of disturbance of vision depends upon the part of the retina which has become detached, and the extent of the detachment. If the macula lutea, or its region is implicated, it obviously follows that the vision is very seriously affected, if not entirely lost.

Distortion of images is also a prominent symptom in detached retina, owing to the unevenness of the retina at the point of separation.

Detachment of the retina is usually seen at the lower portion of the fundus. If the exudate should take place in some other portion of the retina, the tendency is for the fluid to work its way by force of gravitation, between the choroid and retina, to the lower portion of the fundus.

Whatever portion of the retina is detached, the field of vision will be affected upon an opposite part;

for instance, the detachment is on the lower portion
of the fundus, the defect in vision will occur in the
upper portion of the visual field. Should the detachment occur on the nasal side of the fundus, then
the sight suffers on the temporal side of the field of
vision. No matter what part of the fundus is affected, except its central portion, the defect in vision
is always found on an opposite place in the field of
vision.

There is usually great diminution of tension in
retinal detachment. This is caused by a shrinkage
of the vitreous.

Treatment.—Rest in the recumbent posture, and
firm presssure upon the eye, with cotton pads, supported by a bandage.

The iodide of potassium should be given in large
doses, in the manner already heretofore described,
for its absorbent effect. The bowels must be regulated.

Diaphoresis has been recommended by the use
of pilocarpine, but if it is administered, it must be
done with great caution.

Ophthalmoscopic appearances in detached retina:

The detached portion of the retina has the appearance of a grayish membrane, projecting forward
into the vitreous.

The blood vessels can be as readily seen upon
the detached portion as upon the healthy, but they do
not appear continuous, because at the limit of the
healthy retina they bend over and are entirely obscured, when they are again seen upon the detached

portion, because at the limit of the healthy retina they bend over on account of the protuberance of the detached portion, caused by the exudate beneath it.

CHAPTER XII.

DISEASES OF THE OPTIC NERVE.

OPTIC NEURITIS.

As the name indicates, *optic neuritis* signifies an *inflammation of the optic nerve.*

The terms "*papillitis*," "*descending neuritis*," and "*choked disc*," ophthalmically imply the same condition, but as our observation is confined to the optic disc or papilla alone, the term "papillitis" has hitherto been applied; but as the optic nerve cannot usually be affected at the papilla alone, and as papillitis is simply an evidence of the general condition of the nerve, *optic neuritis* is considered more preferable.

As in inflammation of other parts, the papilla is swollen, in some instances to such an extent as to bulge forward into the vitreous, giving it, as described by some authors, a mushroom appearance. This condition originally suggested the term "choked disc."

In inflammation of the optic nerve, the well marked marginal lines of the disc, as seen normally, have entirely disappeared, and in its stead is the "woolly" or hazy appearance.

The central vein is increased in size, and its

branches are distended and tortuous, while the artery is contracted.

In the early stages of inflammatory action the color of the disc is red or livid, owing to the intensity of the congestion; afterward it is changed to a grayish, or grayish-white appearance, in the form of stria extending from the papilla into the surrounding retina.

On account of the extreme congestion of the veins, there are occasionally hemorrhages, marked by flame-shaped or diffused patches, on or near the papilla, although these hemorrhages do occur very frequently in other portions of the fundus.

Strange to say the vision is frequently very little impaired, during the inflammatory stage; however, it is occasionally very much diminished, even when the inflammatory action is not very well marked

In due course of time, optic neuritis reaches its height, after which there is a gradual abatement of all inflammatory action.

The opacity slowly subsides, and if resolution takes place promptly, the papilla becomes perfectly clear, and vision is restored to its normal condition.

This by no means is the termination of all cases of optic neuritis, for it frequently occurs that resolution is not prompt, and the disk presents a more or less blanched appearance, with more or less defect of visual acuity, and occasionally it becomes completely atrophied, with entire loss of vision.

The longer the exudation remains, the more danger is there that the nerve fibres may become permanently injured by the pressure upon them, and the vision more or less affected thereby. The loss of vision in optic neuritis usually comes on slowly, though gradually, but permanently.

Besides the impairment in the acuity of vision, in optic neuritis, the color sense, especially for red or green, is usually interfered with, so much so, that the patient is frequently unable to distinguish them.

The field of vision is also more or less affected, the blind spot being generally enlarged, and the boundary of the visual area contracted.

In optic neuritis both eyes are usually sooner or later affected, but vision is not generally the same in both at the same time.

There are many causes for optic neuritis, but the most common without doubt is intra-cranial disease, notably cerebral tumors; then follow tubercular meningitis and hydrocephalus. Optic neuritis is also observed in albuminuria, glycosuria, lead poisoning, and the abuse of alcohol and tobacco. The loss of vision from the abuse of alcohol and tobacco is very gradual and occurs in both eyes at about the same time, and usually to the same extent.

Treatment.—The treatment of optic neuritis consists in protecting the eyes from the bright light by the use of smoked glasses, and in the removal of the cause of the affection.

Any error of the refraction must be corrected, not that it is by any means the cause of but a very small per cent of the cases of optic neuritis, but it may be an exciting cause, where there is a latent tendency to the affection.

ATROPHY OF THE OPTIC NERVE.

Atrophy of the optic nerve is a result of inflammation, and usually follows in the wake of optic neuritis.

The initial subjective symptom is a diminution of visual acuity, which as a rule proceeds slowly but surely toward total loss of vision.

The progress of atrophy of the optic nerve is not always constant, but is subject occasionally to considerable variation, sometimes becoming stationary, and remaining so for an indefinite time; on the other hand, its progress is, in some cases, very rapid, and vision is entirely lost in a very short time.

The appearance of the optic disk varies in proportion to the amount of the atrophy present. The disk appears very white in advanced cases, and the capillaries are entirely obliterated. The retinal arteries usually appear very small and the veins large and tortuous.

It must be remembered, however, in making a diagnosis, that there is a condition of the normal eye, occasionally seen, in which the disk is very white, and resembles very much an atrophied nerve head. This condition associated with good vision would preclude atrophy.

Beside the loss of vision, impaired color vision, first for green and then for red, is a constant symptom in atrophy of the optic nerve.

There is also more or less contraction of the visual field, as the disease advances to complete atrophy, until all perception of light and color is lost.

The causes of atrophy of the optic nerve are numerous, but it is usually the result of intra-cranial lesion, or spinal affections, the most important of the latter being *tabes dorsalis*.

As in optic neuritis, atrophy of the disk is very often due to syphilis and diabetes, but particularly to the former.

Atrophy of the optic nerve usually attacks both eyes, and the affection usually terminates in complete loss of vision.

Treatment cannot be of great benefit, especially if the affection is due to intra-cranial or spinal lesions. In syphilis, if taken early, constitutional treatment has been effective in staying the affection.

TUMORS OF THE NERVE.

Tumors of the optic nerve and retina are rare. Those of the former usually spring from the sheaths of the optic nerve, and destroy vision by their pressure upon the nerve fibres.

Glioma of the retina is mostly found in children soon after birth. It is usually observed from its characteristic yellowish-gray reflection from the fundus of the eye.

Glioma of the retina is an extremely malignant affection, and the removal of the eye is indicated at the earliest possible moment in order to save the patient's life.   Even this is a *dernier ressort*.   A relapse is almost sure to occur, the disease extending to the brain through the optic nerve, leading to rapid death.

CHAPTER XIII.

GLAUCOMA.

GLAUCOMA is a disease characterized by an increase of the intra-ocular pressure, caused by an augmentation of the humors of the eye, principally of the vitreous humor.

If there is an increase of the tension of the eye, due to an augmentation of the aqueous humor, we find the anterior chamber very deep. This condition is usually due to an iritis, and, although not recognized as glaucoma, is, nevertheless, *increased intra-ocular pressure.*

On the other hand, if there is an augmentation of the vitreous humor, then the anterior chamber is very small, the iris being pushed forward toward the cornea by this increase from the enlargement of the vitreous chamber. This condition is regarded as glaucoma proper.

The theory most generally accepted in regard to an augmentation of the vitreous humor is this: There is a serous exudation from the vessels of the ciliary body and the choroid; by the process known as endosmosis the exudation passes through the hyaloid membrane into the vitreous cavity.

All of the exudate may not pass through into the vitreous, but it serves to increase the pressure from within just as much as if it were within the hyaloid membrane.

The principal exit of the fluid from the vitreous chamber is over the lens and through the pupil into the anterior chamber. Because of the sudden and abnormal increase of the exudate into the vitreous, the iris is pushed forward, and thus the outflow which takes place through the meshes of the ligamentum pectinatum is entirely obstructed, or very much interrupted in reaching the canal of Schlemm.

Upon this theory, the ciliary body and choroid, as well as the ligamentum pectinatum, the canal of Schlemm, and the canal of Fontana, have important relations with intra-ocular pressure.

For the purpose of intelligent study, glaucoma may be considered under the following heads:

$$\text{Glaucoma} \begin{cases} \text{Primary} \begin{cases} \text{Acute} \\ \text{Sub-acute} \\ \text{Chronic} \end{cases} \\ \\ \text{Secondary} \end{cases}$$

Primary glaucoma is an increase of intra-ocular pressure, for which we can discover no reason, and is, therefore, presumed to be independent of any other disease of the eye.

Secondary glaucoma is an increase of the intra-ocular pressure, the result of some other disease.

Primary glaucoma is divided into the *acute*, *sub-acute*, and the *chronic*.

Acute glaucoma is a sudden increase of intra-ocular pressure accompanied by severe inflammatory action. This is the *glaucoma fulminans*, and is especially violent and rapidly destructive in its character, the eye being often entirely lost in a few hours.

The causes of acute glaucoma are obscure, but the consensus of opinion now is, that the ciliary body and the choroid, particularly the former, are primarily participants in the production of the exudate, which augments the intra-ocular pressure, in all forms of glaucoma. This opinion is well supported from this fact, that hyperopic eyes are predisposed to this affection, while myopic are not.

In glaucoma fulminans, the prognosis is very unfavorable; but if an iridectomy is made at once, the eye may be preserved with fair vision.

If, however, the operation must, for some reason, be deferred, a few drops of a one grain solution of eserine should be instilled in the eye every half hour, together with the application to the eye of cloths wrung out of warm water as hot as can be possibly borne, and changed as often as they become cool.

It must be remembered that this treatment alone will not suffice, no matter if there appears to be some relief. *Iridectomy is the sheet anchor in glaucoma*, no matter what its form, but particularly in the acute variety, and the more promptly it is made, the sooner will the patient be placed in a position of safety.

Sub-acute glaucoma is recognized by a less se-

vere inflammatory action and to periodicities of entire freedom.   In this form of glaucoma, there is only
slight injection of the conjunctiva, which is confined
to the circumcorneal zone.

As there is great danger of the sub-acute form
merging into the acute, a diagnosis should be arrived
at as early as possible, that proper treatment may be
promptly applied.   As before stated, iridectomy is
the refuge of safety.

Chronic glaucoma is particularly distinguished
from the acute and the sub-acute forms by the entire
absence of all conjunctival irritation.   This absence
of inflammatory symptoms occasioned this form of
glaucoma to be classed as a non-inflammatory affection.

In chronic glaucoma the exudation takes place
so subtly and insidiously, that the parts affected
become gradually adapted to the condition without
seeming inflammatory reaction.   For this reason patients frequently lose their vision without apprehension of the gravity of the affection, their physicians often thinking that the defect in vision is
due to the formation of cataract.

The treatment of chronic glaucoma is the same
as in the other forms, an iridectomy as soon as the
diagnosis is confirmed.

Secondary glaucoma is an increased intra-ocular tension, due to some other affection of the eye,
the most common of which are complete posterior
synechia, foreign body within the eye, dislocation

of the lens, perforating corneal ulcer with intrusion of the iris, and wounds and injuries of the globe involving the lens, ciliary body and the choroid.

GENERAL CONSIDERATION OF GLAUCOMA.

I believe there is no one thing that would augment the humanity of the physician so much as to make a searching study of glaucoma, sufficiently, at least, that he may be able to warn his patient of the impending danger, and point him, if nothing more, to a road of safety.

As the different varieties of glaucoma often become so blended together, that it is difficult to differentiate one from another, or even to diagnose glaucoma itself from some other affection, I will endeavor to point out the most important symptoms, and give the distinguishing characteristics of this affection as compared with some other affections, so that the observer may more readily recognize this disease.

There are three affections with which this disease may, by the casual observer, be confounded, viz: *neuralgia, iritis*, and *senile cataract*.

Acute glaucoma may be, and often is, mistaken for neuralgia—the pain in and around the eye often simulating that affection.

The periodicity of sub-acute glaucoma, in which there is entire freedom from pain, whilst this is one of the landmarks in its diagnosis, is also diagnostic of neuralgia.

In iritis there is always an increase in the intra-
ocular tension, but upon close inspection we find
that the anterior chamber is abnormally deep,
whereas in glaucoma it is very shallow.

It was formerly the general opinion of the physician that gradually increasing blindness without
pain, especially in the aged, was due to the formation of senile cataract. For this reason many persons were permitted to lose their vision, beyond redemption, before an examination revealed the fact
that they were suffering from glaucoma.

In order to have a more methodical arrangement
of the subject, we will consider it under the following
subdivisions, viz: the *subjective* and the *objective
symptoms.*

SUBJECTIVE SYMPTOMS.

In acute glaucoma the attack is usually at night.
The pain accompanying it is violent, radiating from
the eye through the head. There is also considerable
conjunctival irritation.

The first and most important symptom in sub-
acute glaucoma is pain, with intermissions of entire
freedom. With each paroxysm there is some appreciable loss of vision, which is never recovered. The
paroxysms are more severe at night.

In hemicrania or megrim, on the other hand, the
pain is more violent during the day, commencing,
often, very early in the morning.

The intermission may be a few days, or it may be
weeks, or a month, or as long as a year.

We will find upon inquiry that the patient has had considerable uneasiness, and frequently much pain of a neuralgic character about the supra-orbital and temporal regions, and along the side of the nose.

In chronic glaucoma there is a gradual loss of vision, but as a rule it is not accompanied by pain, although when the patient is questioned, it is frequently ascertained that there has been a sense of "uneasiness" about the eyes.

Where the increase of tension is so slow, there seems to be a gradual adaptation to the condition; hence the absence of pain.

It is in the chronic form that so many are deceived, the sight being so gradually lost, that it is often mistaken for the formation of cataract.

A very important symptom in any form of glaucoma is the appearance of the colored halo around the flame of a lamp or candle.

OBJECTIVE SYMPTOMS.

The old are more subject to glaucoma; and for this reason it is thought that the strain upon the ciliary muscle in the act of accommodation may have much to do with precipitating this affection.

There is increase of tension; that is, increased hardness of the eyeball.

In the earlier stages there is apparently no increase of tension; as the disease advances, however, there is increased tension, sometimes not very marked,—in other cases the eye is of a stony hardness.

There is diminished depth of the anterior chamber, and the iris and lens are pushed forward toward the cornea, in advanced cases almost touching it.

The cornea often appears semi-translucent, having lost its brightness.

There is dilatation with sluggishness of the pupil, that is, the pupil responds very slowly, if at all, to the influence of light.

In the case of neuralgia, there is contraction of the pupil. In cataract there is dilatation of the pupil; but when uncomplicated, the pupil responds readily.

The cupping of the optic papilla, together with the pulsation of the arteries of the same, one of the most important symptoms in glaucoma, can only be detected with the ophthalmoscope.

With this information, I believe the careful physician of this day will not fail to diagnose a case of glaucoma, and by the timely application of the proper remedy, rescue his patient from an impending blindness.

CHAPTER XIV.

DISASES OF THE VITREOUS HUMOR.

Hyalitis.

HYALITIS is an inflammation of the vitreous body. It often follows as the result of inflammatory changes in adjacent structures. It may also occur after the entrance of a foreign body into the vitreous.

There are two forms of hyalitis, the *suppurative* and the *non-suppurative*.

In the *suppurative form* there is a collection of pus in the vitreous, and the fundus may be partially or completely obscured.

Timely enucleation should be resorted to when pus has once formed in the vitreous, for the reason that the other eye may become, through sympathy, involved.

In the *non-suppurative form* there is the formation of opacities in the vitreous, usually the result of a retinal or choroidal hemorrhage.

Opacities of the Vitreous.

Opacities of the vitreous usually appear as black spots floating before the eyes. They are usually brought to the patient's attention by looking at a

TEXT-BOOK OF OPHTHALMOLOGY.

white surface, when they appear as dark spots upon the white background.

They vary in shape, appearing as threads, flakes, dots, and membranes before the eyes and may be fixed or movable.

They may be readily observed with the ophthalmoscope, when they appear as floating black particles, or clouds before the illuminated fundus.

Causes.—Vitreous opacities may occur as the result of a hemorrhage from the choroidal, retinal, or ciliary vessels which may follow an injury to the eye or an inflammation of the retina, choroid, or ciliary body.

They may also occur as the result of errors of refraction, according to DeSchweinitz.

They may be found in connection with constitutional diseases, as anæmia and syphilis, and are frequently the result of chronic constipation.

Prognosis.—In vitreous opacities, the result of suppuration, the prognosis is decidedly unfavorable.

In the young and middle-aged, where the opacities are due to hemorrhage into the vitreous, and are not of an extensive character, they are usually absorbed, and the prognosis is favorable; but in old persons the opacities usually remain through life and cause more or less annoyance to the patient.

Where the hemorrhage is very large, the clot may become organized and detachment of the retina follows.

In cases of syphilis or other constitutional disease, the results obtained by the use of the proper remedies for these affections are most favorable.

Treatment.—Very favorable results can usually be obtained by the proper use of pilocarpine to produce sweating.

Where the opacities are due to syphilis, the proper remedies for that disease produce the best results.

Where anæmia is the cause, iron and other tonics should be administered.

The bowels should be kept open when there is a tendency to constipation, and in cases where there are extensive hemorrhages, local blood letting should be resorted to.

Any error of refraction should be corrected, and the patient warned not to strain.

Patients having vitreous opacities often become greatly depressed, fearing total blindness as a result. It is within the province of the physician to encourage such patients as much as possible with the assurance of the fact, that these opacities will not probably increase, and he should instruct them to endeavor to disregard the floating specks.

SYNCHISIS.

The vitreous humor occasionally, on account of an inflammation of the retina, choroid, or ciliary body, loses its normal solidity and becomes softened or fluid in consistency, which condition is known as *synchisis*. This condition is easily determined with the ophthalmoscope by the rapidity with which the opacities move when the eye is turned in different directions.

There is also a condition known as *scintillans synchisis*, or *sparkling synchisis*, accompanying softening of the vitreous, caused by small flakes of cholesterin, which have, with the ophthalmoscope, the appearance of fine gold dust floating in the vitreous.

PERSISTENT HYALOID ARTERY.

A dark, thread-like substance has been observed running part or all the way between the optic disc and the posterior pole of the lens. This is a remnant of the hyaloid artery which existed in intra-uterine life, and which has not become completely absorbed.

CHAPTER XV.

CATARACT.

THE term cataract signifies an opacity of the crystalline lens, or of its capsule, or of both the lens and its capsule together.

Opacity of the crystalline lens is known as *lenticular cataract*.

Opacity of the capsule is denominated *capsular cataract*.

When both the crystalline lens and its capsule are opaque, it is termed *capsulo-lenticular cataract*.

Clinically, cataract may be considered under the following forms:

	senile	{ cortical	
		nuclear	
Cataract	congenital	{ complete or partial	{ lenticular usually anterior polar and posterior polar
	traumatic		
	secondary		

Senile cataract is probably due to a degeneration of the lens fibres, in consequence of defective nutrition. It occurs, usually, after forty-five years of age. The lens in this case is usually hard.

Senile cataract may be *nuclear* or *cortical*.

Nuclear cataract is that form of opacity of the lens which commences in the central portion of that body.

Cortical cataract is an opacity of both surfaces of the lens. This form of cataract appears as opaque stria, in the form of prisms, with their bases directed outward, towards the periphery of the lens, and the apices toward the center. As the periphery becomes opaque first, this form of cataract is frequently denominated peripheral cataract.

Congenital cataract is a form of cataract which is acquired in intra-uterine life, and exists at birth.

Congenital cataract may be *partial* or *complete*.

Partial cataract of the congenital form consists of two varieties, *anterior polar cataract*, and *posterior polar cataract*.

Anterior polar or *pyramidal cataract* consists of an opacity of the central anterior portion of the lens capsule. This form of cataract is not always congenital, but is often due to a central perforating ulcer of the cornea in ophthalmia neonatorum, wherein, during the absence of the aqueous humor, the anterior surface of the lens capsule comes into contact with the posterior surface of the cornea; at this point it is not unusual for the parts to become opaque.

Posterior polar cataract is a central opacity of the posterior portion of the lens capsule. This form of opacity is generally small and round, but sometimes it is stellated, or has small streaks extending from it, making its edges appear fringed.

Posterior polar cataract is probably due to an imperfect absorption of the fœtal hyaloid artery.

In the complete congenital opacity, the lens is always soft.

Traumatic cataract is an opacity of the lens or its capsule caused by an injury of the eye. In this case the lens capsule is usually ruptured, and the aqueous humor is permitted to come into contact with the lens substance.

Secondary cataract consists of portions of the lens or its capsule, which have been left behind, and have become opaque after the cataract has been extracted.

There are other forms of cataract, not specially classified here, as the *diabetic*, due to disturbed nutrition, and the *complicated*, due to disease of other parts of the eye.

Experience has fully demonstrated, at this day and age, that there is only one operation for hard cataract deserving consideration, and that is extraction, of which the following may be considered as constituting the accepted methods for accomplishing this purpose.

EXTRACTION OF CATARACT, DIFFERENT METHODS.

Since the first operation for the extraction of cataract, there have been so many methods devised for its performance, and each one has been so tenaciously supported by its originator, that the beginner is bewildered as to which particular one to adopt. Like surgical operations in general, no matter what particular method is employed, nor how well the operation is performed—however favorable the condition of the patient—failures come, and frequently when least expected; hence the impossiblity of a procedure that is invariably successful. No doubt each method for the performance of the operation has its particular advantages as well as its peculiar dangers. Consequently, a searching study of these various methods should be encouraged, adopting the operation or that part of the operation which has been proven by experience to be advantageous-—discarding that which has been shown to be hazardous and dangerous.

By many ophthalmologists it is considered that there are but two methods of extraction: The flap operation and the linear extraction—the others being merely modifications of these two. However, for the purpose of discussing this subject in a practical manner, it will be necessary to take into consideration the following as distinct methods: First, The Flap Operation. Second, The Linear Extraction, Third, The Modified Linear Extraction, or Von Graefe's method. Fourth, Pagenstecher's method. Fifth, Lebrun's method. Sixth, MacNamara's method. Seventh, Bell Taylor's method.

The Flap Operation.—Of the different procedures for the object named, there has been none whose result, when a success, is all that could be desired,

so much as in this operation. When everything does
well, the result is almost perfect; the patient has a
central, movable pupil, and the wound heals with
such exact apposition that the normal curvature of
the cornea is not in the least impaired; but so many
accidents are liable to happen, and the after-treat-
ment is so troublesome, that the operation has fallen
into disrepute.

Among the many objections which are urged
against it are: the tendency to suppuration of the
cornea, the vigor of the patient generally being unequal to so large an area of cut cornea; the frequency
of a prolapse of the iris—forcing the edges of the
wound apart; the liability of a reflection of the flap,
on account of a slight movement of the eye or the
lid after the eye is closed, and before the bandage
is applied; imperfect coaptation of the lips of the
wound, then there is not union by first intention,
but by granulation—thus, the normal curvature of
the cornea is interfered with, and, as a result, astigmatism.

The Linear Operation.— This operation having
undergone so very many modifications, it is now
hardly recognized under its old name; it is, in fact,
generally described as the *Traction Method.* As described under the head of Gibson's Operation, it consists in dilating the pupil and lacerating the lens
capsule with a needle, as if operating for solution—
only the capsule must be more freely incised. A
few days after the needle operation, an incision is

made in the sclero-corneal border, at its upper part, including about one-fourth of the corneal circumference; a curette is introduced into the anterior chamber and turned edgeways, so as to open the wound in the cornea; then by slight manipulation on the lower part of the cornea, the lenticular matter is allowed to escape.   No iridectomy is made.

This operation was attended with considerable danger, in consequence of the swelling of the lens, which from the pressure exerted by it caused considerable irritation.

Von Graefe's Modified Linear Extraction.— This operation, sometimes called the *Peripheral Linear Extraction*, was introduced by Von Graefe in 1865. The incision lies considerably beyond the sclero-corneal junction, involving the conjunctiva, of which a flap is made.   As recommended by its author, the iridectomy should be made some weeks before the extraction; however, the iridectomy is now almost always made at the time of the extraction.   The plain points in the operation are simply these: The incision must be made entirely within the sclerotic— beyond the sclero-corneal border, and an iridectomy must always be performed.   The merits of this method were much lauded, and it was confidently hoped that the objections which were attendant upon the flap operation would be overcome.   This was in a great degree realized, the advantages lying in the small risk of suppurative inflammation.   But with the new advantages came new and intricate dangers. The near proximity of the incision to the ciliary region caused sympathetic ophthalmia with destructive irido-choroiditis, which was liable to attack and cause the loss of the other eye. There is also great danger of losing the eye by carrying the incision too far into the sclerotic and rupturing the hyaloid membrane, causing an escape of the vitreous humor. Another great disadvantage is the tendency to hemorrhage from the conjunctiva into the anterior chamber, thus obscuring the parts and making further procedure difficult. Even though the operation should be successful, there is another important objection; the sight is imperfect and confused by circles of diffusion on the retina, which is always the case when there is a loss of any part of the iris.

Prof. Pagenstecher's Method differs very little from the flap operation. The incision is made as nearly as possible at the sclero-corneal margin with a Graefe's knife; a large iridectomy is also made. By gentle pressure the lens within its capsule is forced out. If the lens is not readily displaced in this way, then a curette is inserted behind the lens, and by gentle traction exerted on it, it is started from its position. The incision is made downwards.

This method, no doubt, is a valuable one, for where the capsule is removed with the lens, there can be no obstruction of vision from portions of it remaining behind, which frequently happens when the lens is removed in the ordinary way—that is, by

rupturing the capsule. The passage of the curette
behind the lens is dangerous practice, frequently
causing a rupture in the hyaloid membrane and a
loss of the vitreous. The eye is also subject to the
same dangers as in the ordinary flap operation.

Lebrun's Method makes the incision entirely
within the cornea. The knife is inserted opposite
the centre of the pupil, about midway between the
centre of the cornea and the sclero-corneal margin;
it is carried directly upwards or downwards—according as it is to be an upper or lower flap—on this
same line, and consequently the knife will emerge
at a point equi-distant from the centre of the cornea
and the sclero-corneal margin. Iridectomy is not
performed, and, after the lens is extracted, eserine
is employed to prevent anterior synechia.

A moment's reflection demonstrates its peculiar
advantages, as well as its disadvantages. The corneal wound being so far from the ciliary body, there
is little danger of cyclitis; there is not likely to be
a loss of the vitreous, nor a wounding of the iris
whilst making the section. If there is union by first
intention, then, as in the flap operation, we have all
that can be desired. However, like the flap operation, it is liable to the same dangers—suppuration
of the cornea; a reflection of the flap and imperfect
coaptation of the lips of the wound. I am very sure
that, in this method, there is more danger of a reflection of the flap than in the flap operation; for,
in this case, the flap is thinner and has not the sup-

port within itself to sustain it, as has the old flap. The wound is extensive and the healing process necessarily slow, especially in persons of feeble constitutions. The object in making the wound so far out in the cornea, together with the use of eserine, is to prevent a staphyloma and an adhesion of the iris in the lips of the wound. The use of a myotic in such cases has been discarded by every observant ophthalmologist who has had experience with it. If there ever is a time that myosis should be avoided, it certainly is immediately after the lens has been pressed through the pupil—for the pupillary margin is certainly liable to be bruised, and its tendency is to the promotion of plastic lymph, and as a result occlusion of the pupil.

Macnamara's Method consists in the extraction of the lens through an incision made at the extreme outer margin of the cornea. The lens is removed by the scoop. The section is made with a broad bladed, triangular knife, whose point is thrust through the line of junction of the cornea and sclerotic, as before stated, on the temporal side of the eye. The author says, "the blade of the knife is to be passed steadily onwards, nearly up to its heel, so that the incision made through the sclerotic is at least half an inch long." An iridectomy is not made unless the pupil does not dilate; and his aim is to remove the lens in its capsule, particularly if there is much transparent, cortical matter round the lens. If there is much traction necessary in removing the lens

in its capsule, then he ruptures the capsule and re-
moves the lens in the usual way.

This operation, although subject to some of the
dangers attending the flap operation, is far more
practical. Like the flap operation there is danger
of suppurative inflammation, but the author claims
he seldom lost an eye in this way. There is certainly very little danger from a reflection of the flap,
as the movements of the eye or the lids would have
very little influence in separating the lips of the
wound.

Bell Taylor's Operation consisted in removing the
lens through a peripheral opening of the iris. The
incision is made in about the same manner as in the
linear method, and the lens is extruded through a
gape in the iris, gliding behind the pupil. The author claimed "all the advantages in the way of cer-
tainty and safety of an associated iridectomy, and
the grand desideratum—a central and moveable
pupil."

The result in successful cases was almost perfect, but the operation was found to be too complicated, much time and great dexterity being required
to procure the passage of the lens through the small
opening in the iris. The upper portion of the pupillary margin did not always preserve its arc; but,
drooping at a point corresponding with the incision
in the iris, formed a small and irregular pupil. The
operation has, I believe, been abandoned by its
author.

Now, from an examination of these procedures, we have learned that the sources of danger are numerous, and frequently in avoiding one, we are liable to fall into another.

The following may be taken as a summary of some important facts: That in making an incision we must not encroach too closely upon the ciliary body—thus avoiding destructive cyclitis and sympathetic ophthalmia, as well as increased liability of rupturing the hyaloid and loss of the vitreous.

That in flap operations there is danger of suppuration, and that the danger increases according as the incision is large and the flap long and unable to support itself.

That the nearer the incision is made to the corneal margin, or sclero-corneal junction, the greater the danger of a staphyloma of the iris.

That the retention of the entire iris with a central movable pupil gives the greatest possible degree of success.

That the wounding of the iris is to be deprecated—unless in a case of exigency, and then as a *dernier ressort*.

That the use of myotics after extraction is detrimental to a favorable result and should be avoided—but, on the contrary, the use of mydriatics gives great assurance of the avoidance of dangerous complications.

That the use of the scoop for the extraction of the lens is hazardous and should not be resorted to except in an emergency.   The danger lies in rupturing the hyaloid.

That the use of the speculum, if possible, should be avoided for the reason that, when the incision is made, the aqueous humor frequently escapes with such rapidity that the hyaloid membrane is ruptured, on account of its sudden loss of support, and the vitreous escapes, frequently before the speculum can be removed and the eyelids closed.

That wounds of the cornea, however slight the puncture or scratch, are not devoid of risk, but the danger is confined to those who are not healthy. On the other hand, if the patient is healthy the less disposed is the eye to take on suppurative inflammation.   The most extensive cuts and lacerations of the cornea recover with surprising activity, even in advanced age.

From a careful resume of the preceding, I have endeavored to select a method for extraction which will possess the following points:

An incision that is small and entirely within the cornea, and so made that there is no flap, and no gaping of the wound.

Almost entire painlessness, the avoidance of an anesthetic, the speculum and fixation forceps.

A central movable pupil, except in very rare cases, where the pupil will not dilate sufficiently to allow the passage of the lens.

Wright's Operation.—As the principal lesion in this particular operation is in the cornea, and as this is the point wherein this process differs most in importance from all other methods of removing the lens, an occasion presents itself which invites our consideration of some important facts concerning wounds of the cornea in general.

It is well known that there is no scratch or abrasion or puncture of the cornea, however small or seemingly insignificant, which is devoid of risk. On the other hand, there is no wound within certain bounds from which the eye may not recover under favorable conditions, the most important of which is the vigorous and healthy condition of the patient, which promotes in the greatest degree the restorative powers.

It is generally the case that in wounds of the cornea where it is abraded and a considerable portion of the epithelium is lost, it is more liable to take on suppurative inflammation than from an incised wound.

Considering that accidental wounds of the cornea are mostly inflicted by blunt instruments—pieces of machinery or of stone, glass, wood, or belting—the wounds are more frequently lacerated than incised, the epithelium often extensively abraded, the eye itself frequently injured from the concussion and force of pressure from the inflicting body, yet with all these complications it is surprising how very rapidly it heals, what slight inflammation exists after such severe and extensive injuries, and how small the cicatrices resulting from large and intricate wounds. Where vision is lost in these cases it is not frequently because of opacity of the cornea, but mostly from occlusion of the pupil, because it was not kept under the influence of a mydriatic, or else from opacity of the lens as a result of the concussion the eye received at the time of the injury.

How much more rapidly then should these wounds heal when made by a clean, sharp instrument, in the absence of a heavy shock or concussion, and of the dirt and extraneous matter usually present in cases of accident!

But it is not so with the sclerotic. It cannot be wounded with such impunity. When this structure is injured, however simple and small the wound, the danger is great and the eye is almost always lost. The reasons for this are few and simple. It is apt to be complicated with a prolapse of the choroid and ciliary body into the wound, and as it cicatrizes, sympathetic irido-choroiditis is established in the other eye, and we are necessitated to remove the wounded eye in order to save the other.

The nearer the choroid, ciliary body and retina are approached, either designedly or accidentally, the greater the danger of losing the eye from the resulting inflammation.

Taking these facts into consideration, I have attempted to obtain the highest possible degree of success, by making the necessary wound in a part as remote as possible from the seat of danger and where the tendency is to heal rapidly—avoiding that structure which is most liable to dangerous complications; hence I have adopted the plan of making the incision entirely within the cornea, and at a distance as remote as possible from the sclerotic.

Having our patient ready, we will then proceed to the operation, first by dilating the pupils. I usually drop a 4 per cent. solution of cocaine in both eyes every minute, for ten minutes. The pupils will dilate much better if the solution is dropped in both at the same time.

If at the end of ten minutes the pupils have not become well dilated, a 1 per cent solution of atropine should be dropped into the eye every five minutes for half an hour; at the same time the cocaine solution should be continued sufficiently often to keep up its anesthetic effect.

If the pupil has dilated well in the eye upon which we desire to operate, we will not make an iridectomy; however, if to the contrary it has not dilated well, we must be very sure to make an iridectomy. This, I claim, is an important point. If the pupil appears sluggish and does not dilate fully and promptly, then it is very important that we make an iridectomy, for by so doing we will prevent an occlusion of the pupil.

Where the pupil dilates freely and promptly, there is no danger of iritis or an occluded pupil, unless the anterior chamber is filled with lenticular matter and fragments of the capsule.

The Operation.

For convenience we imagine a transverse horizontal line one-third the way down from the upper edge of the cornea. A medium-sized Graefe's knife is used in making the incision. The point of entrance must be near the sclero-corneal border, but entirely within the cornea, on the imaginary line, as is shown in the dotted line in Fig. 24.

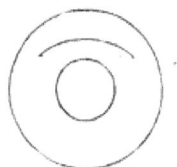

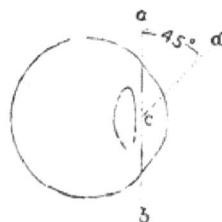

FIGURE 24.   FIGURE 25.

The cutting edge of the knife is directed upwards and forwards, the plane of the knife making an angle of about 45 degrees with the plane of the normal iris, as is shown in Fig. 25. (The line *a b* represents the plane of the normal iris; *c d* the plane of the knife).

The back of the knife must be kept on the imaginary line and the counter-puncture made near the sclero-corneal border at a point as nearly as possible corresponding to the point of entrance. The knife is now steadily and gently pushed forward until the section is completed. The line of incision will now present the appearance of the curved line in Fig. 24.

The patient is now allowed to rest a moment when, if the pupil had not dilated well, the iridectomy is made. On the other hand—as has been indicated before—if the pupil had been fully dilated before the incision was made, the iridectomy is to be omitted.

The next step in the operation, as usually practiced, whether an iridectomy has been made or not, has been to rupture the capsule of the lens; but I am fully satisfied that as a rule the lens is easily removed within its capsule, and especially so in this particular operation.

Dr. Macnamara in commending this procedure says: "The advantages it offers are, that no capsular cataract can possibly form, and there is no chance of any soft lenticular matter being left clinging to the iris and setting up inflammation in that delicate structure; and the greater my experience in these matters, the more convinced I am that most of our failures in extraction are due to the fact of soft lenticular matter and capsule being left in the eye after the removal of the lens."

I am disposed to this opinion, that there would be very few cases of iritis as a result of these operations were it not for these portions of lenticular substnce and capsule which are allowed to remain, and which it is frequently impossible to remove, because of the danger in losing the vitreous.

We will then without rupturing the capsule take the next step in the operation, which is to press down

the upper segment of the cornea with the finger as
is represented in Fig. 26. At the same time, if the
lens does not readily present itself, slight pressure
with a finger of the other hand may be made near the
sclero-corneal border below, which is also shown in
Fig. 26.

If now the lens does not readily present itself
and pass out, we may rupture the capsule, and again
proceed to make pressure as has been directed. The
lens will now present itself and pass through the
opening as represented in Fig. 26.

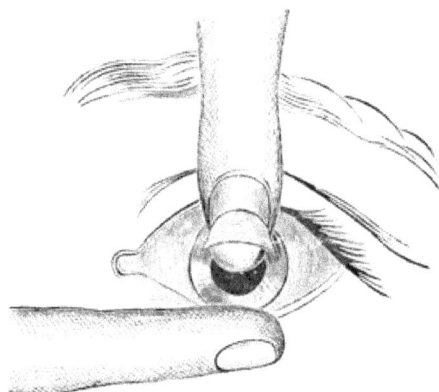

FIGURE 26.

I have found this procedure much better and
safer than by making the pressure with a curette or
scoop, as was formerly recommended. The surgeon
who once uses his fingers in the manner here described, will doubtless never thereafter resort to an
instrument for the purpose of making pressure. In
this way the eye can be handled with much more

safety and comfort to the patient, and with greater facility to the operator.

I rarely am compelled to make pressure below the incision at the sclero-corneal border, more than to steady the eye whilst the pressure is being made on the upper segment—for the lens usually passes out as soon as the pressure above is sufficient to make the opening large enough to allow its passage.

The advantages gained by this procedure are evident.

By other methods sufficient pressure is made below the incision, to make the wound gape enough to allow the lens to pass, frequently rupturing the hyaloid mebrane and causing a loss of the vitreous— whereas, in this method, the opening that is made by pressing down the upper segment, as indicated, is sufficiently large—if the incision has been properly made—to allow the ready passage of the lens.

This mode of operation presents other manifest advantages:

It is the smallest possible incision that can be made, that will allow the ready passage of the lens.

The incision is made in that structure which is most likely to heal and least liable to take on dangerous complications.

The manner in which the incision is made prevents its gaping, hence there is no danger of a staphyloma of the iris.

There being no flap, and the incised surface being small, danger of suppuration of the cornea is reduced to a minimum.

Even if the wound in the cornea should leave an opaque cicatrice, it is covered by the upper lid, and does not interfere with vision in the least, but as a rule the cicatrice clears up perfectly, and in a short time cannot be detected with the naked eye.

On account of the pressure on the upper segment of the cornea, the corresponding portion of the iris is so pressed upon that the lens is brought immediately to its point of exit through the pupil, thus preventing its slipping up behind the iris, and bruising the ciliary body, with the resulting consequences.

Acting upon the principle of not removing any structure which can be retained without interfering with the result of the operation, in most cases we retain the entire iris, and thus secure the advantages of a central movable pupil.

The operation being almost painless, is reduced to the greatest simplicity, not requiring an anesthetic, speculum, fixation forceps, and frequently but one instrument—a Graefe's knife.

In this, as in all other important operations upon the eye, thorough asepsis is of paramount importance; therefore all instruments, dressings and bandages should be thoroughly sterilized, in the most approved manner.

Cutting instruments should not be boiled, nor immersed in the bichloride of mercury solution to render them aseptic, for the former will destroy the temper, and the latter will cause them to rust. If the instruments have been thoroughly washed and dried since having been used, it is quite sufficient to immerse them in absolute alcohol when you operate.

Immediately before the operation, the eye should be thoroughly flushed with a 1-1000 bichloride solution of mercury; this should also be done immediately after the operation is completed, when the lids should be dried and the dressings applied.

Before the dressings are applied, we should be very sure that all extraneous matter is removed from the anterior chamber, that no part of the iris is intruded into the corneal wound, and that the lips of the wound are perfectly adjusted.

The dressings should consist of oval pieces of sterilized cotton wool, sufficiently large to cover the eye, and held in position by a roller bandage, applied with enough firmness only, to give the eye sufficient protection and support. This is an important matter; if the eye is too tightly bandaged, it is liable to cause the wound to gap.

It is necessary to cover both eyes, for the reason if one eye is exposed, and the patient can use it, the bandaged eye will move with it whenever the exposed eye moves.

My rule has been not to annoy the patient by making an examination of the eye until the third day after the operation, unless he is suffering pain or some inconvenience, in which case the eye is examined sooner. But on the third day the bandages are removed, the secretions are washed away from the edges of the lids in both eyes, the eye operated upon is examined, and if the cornea is clear, and the iris and pupil easily seen, we may be reasonably sure of success.

If on the third day everything is favorable, there is very little danger of any trouble thereafter, if the patient will take the proper care of himself. He must keep his bed and talk as little as possible for a few days yet.   After the fifth or sixth day after the operation he may be allowed to sit up a few hours every day, and in ten days, if everything goes well, he may be allowed some outdoor exercise.

Contingencies liable to occur in cataract extraction:

Loss of vitreous.

Lenticular matter in anterior chamber.

Iris caught in wound.

Suppuration of cornea.

1.   *Loss of vitreous.*

While the loss of a small amount of the vitreous during the operation is not usually a matter of very great consequence, yet if a considerable amount is lost, it is of serious importance, for the reason that the retina, having suddenly lost its support, becomes detached.

If at the moment the lens passes through the opening a sudden gush of the vitreous follows, the eyes must be closed at once, and the patient cautioned not to move them and to keep perfectly quiet until a compress, secured by a bandage, is applied over both eyes.

Great caution must be observed in making pressure upon the eye in the attempt to force the lens through a small opening. When it is found that the opening is too small, we should enlarge it rather than to take the risk of rupturing the hyaloid.

2. *Lenticular matter in the anterior chamber.*

Frequently in removing cataract, a considerable amount of cortical substance remains in the anterior chamber, and it is a matter of great importance that it should be removed, as its presence is very liable to excite an inflammatory action which would compromise a satisfactory result.

It often occurs that the cortical substance is clear, while the nuclear portion of the lens is opaque, in which event it is not readily detected. Should the eye be dressed without its removal, the surgeon will be surprised at his next visit to find the pupillary space filled with lenticular matter, which has become opaque from the action of the aqueous humor upon it. Nothing can be done now except to keep the pupil well dilated and its edges as far from the extraneous matter as possible until it absorbs.

If much lenticular matter is allowed to remain in the eye, it is very liable, even in our greatest effort to keep the pupil dilated, to cause an occlusion of the pupil, besides a long course of inflammatory action in the eye, involving the choroid and retina to such an extent as to render futile an operation for artificial pupil, with the contemplation of useful vision as a result.

The eye therefore should be examined very carefully before it is dressed, and should the pupil not appear perfectly clear, gentle stroking with the spatula upon the cornea, will cause the cortical substance, or any loose lenticular matter within the anterior chamber to pass out through the lips of the wound.

3. *Iris caught in the wound.*

If any portion of the iris protrudes, or is caught within the lips of the wound, it should be carefully replaced. This is readily accomplished by a slight rotary motion made by the finger over the closed upper lid.

If at our first examination of the eye, on the second or third day after the operation, we find the iris is protruded through the lips of the wound, an attempt to replace it would be futile, and we can have no preference in the matter but to snip it off as near the cornea as possible. This procedure should be observed at each dressing thereafter, as long as any portion of it projects from the wound.

4. *Suppuration of the cornea.*

Suppuration of the cornea is one of the worst after complications which can possibly follow in the wake of an operation for the extraction of cataract, and once precipitated, the eye is lost beyond redemption.

Suppuration may come on so insiduously and without any pronounced symptoms, that we are not cognizant but what the eye is doing well, until we make an examination preparatory to dressing it, when we are surprised to find the cornea hazy, the edges of the wound separated, presenting a grayish-yellow appearance, and other evidences indicating the formation of pus.

The suppuration however may, and often is, attended by violent pain in the eye and over the supra-orbital region, with oedema of the lids and swelling and oedema of the conjunctiva.

Soon a profuse muco-purulent discharge will begin, and all hope of saving the eye, with the prospect of useful vision, is vanished.

We must now set to work at once to check the inflammatory action, else it will extend to the iris, the ciliary body and the choroid, precipitating a pan-ophthalmitis which is certain to destroy the globe.

Prompt action is required. The eye must be flushed with the bichloride solution, 1 to 1000, and hot water fomentations must be applied over the eye and surrounding parts for at least an hour at a time, and repeated two or three times a day, after which warm cotton wool should be placed over the lids, and secured with a loose bandage.

At the same time the patient should have a good nourishing diet, with wine or brandy, if weak and anaemic. If suffering severe pain an anodyne should be administered.

THE OPERATION OF DISCISSION OR NEEDLING.

The operation of *discission* or *needling* is applicable only to soft cataracts, and consists in lacerating the anterior portion of the lens capsule with a cataract needle by introducing the instrument through the edge of the cornea, and turning its point in such a manner as to bring it into contact with the lens capsule, when it is lacerated as before mentioned.

The object of the operation is to bring the aqueous humor, which has great solvent properties, into contact with the lens substance and cause it to gradually absorb.

It is often necessary to repeat this operation, as it rarely is the case that one operation is sufficient.

Preceding the operation the eye should be well dilated with atropine, and if the patient is a child, and there is an uncertainty that it will be quiet, a general anesthetic should be used, as great care must be taken not to injure the iris in the operation. In older persons the cocaine is sufficient.

After the operation the eye should be dressed with a light compress bandage, and the atropine solution should be applied sufficiently often to keep the pupil well dilated and away from the lenticular substance, because of the danger of the iris being caught in the fragments of the lens.

Great care must be taken in this operation, especially at the first needling, not to lacerate the

lens capsule too extensively, for there is danger of
the lens swelling to such an extent as would injure
the eye.

If the swelling of the lens should be very great
after the needling, an incision should be made into
the periphery of the cornea, and the lenticular matter gently pressed out. This procedure is often
obligatory in order to prevent a general suppuration
of the eye.

There is also a method of suction for soft cataract, whereby the lens substance can be drawn out
through the nozzle of an instrument designed for
that purpose.

In this case, a small incision is made in the periphery of the cornea, and the nozzle of the suction
instrument introduced and brought in contact with
the soft particles of the lens lying loose in the anterior chamber, when they are drawn up into the
instrument.

The old method of *couching* or *reclination*,
wherein the lens was dislocated and pushed down
into the vitreous humor, is simply mentioned as a
relic of the earlier methods of operating for cataract.
Daviel, who introduced this method, made himself
immortal by this operation alone, although perhaps
not one patient in twenty received the least benefit
from the operation, but on the other hand, suffered
severely from the violent inflammatory action which
was almost sure to follow.

CHAPTER XVI.

Sympathetic Ophthalmia.

THERE is no subject in the literature of ophthalmology of more importance than sympathetic ophthalmia, from the fact that a very slight, and to outward appearances, a very trifling injury of an eye often causes total loss of vision in its fellow.

Aside from a foreign body within the eye, a wound of the eye-ball involving the ciliary region, together with a low grade of inflammation of the injured eye, will often finally result in entire loss of vision. This inflammatory condition, however slight and transitory, if persistent in its attacks, has been proven very significant, and is the alarm of danger, which should be heeded at once.

We will first consider the subject of *foreign body within the eye.*

It has generally been considered, that so long as the eye is not painful, and the sight is not materially injured, there need be no apprehension from a foreign body within its structure; but when we take into account the fact that a foreign body may remain in the eye for a score of years, and even longer, and then be the cause of severe inflammation and loss of vision in the other eye, we are inclined to look with anxiety upon all such injuries of that organ.

Whether it is conducive to the best results in all cases, to remove the eye, when there is a foreign body within its structure that cannot be seen, is a question which must be answered on the one hand by the number of eyes which would have retained tolerable vision without pain in the injured organ, against the number on the other hand that suffer pain in the damaged eye together with the danger of losing sight in the other, through sympathetic ophthalmia.

When a foreign body is lodged within the anterior chamber, it generally can be seen and removed; and if the cornea or the iris has not been seriously injured, there will be a speedy recovery.

But the foreign body in entering the anterior chamber may wound the iris or cornea in such a manner as to cause it to bleed, and the anterior chamber may become so filled with blood that it is impossible for us to locate the body, or to indicate the result.

In this case the question very fitly arises, *has the foreign body entered the eye?*

Cases are known where the foreign body had penetrated the cornea and entered the anterior chamber, with or without wounding the iris, and then rebounded, leaving the anterior chamber filled with blood.

In the absence of pain and severe marks of ex-

ternal violence, the surgeon would not feel justified
in at once removing what might in the future, if left
alone, make a useful organ of vision.

Should the foreign body lodge within the iris,
it can generally be seen and removed if the aqueous
humor is clear; but the penetrating body may be of
the color of the iris, or it may be small and buried
within its structure. In this case there is danger of
cyclitis and sympathetic ophthalmia, yet the
thoughtful and conscientious surgeon does not feel
disposed to remove the eye at once, and especially
so long as the patient has vision in it, and is not
suffering pain in the injured organ.

Then what must we do? This is a matter of
no small moment, and every complication must be
taken into account. Here we know the iris is
wounded; but has the penetrating body passed
through it? If the humors are clear we examine
the eye with the ophthalmoscope; we can discover
no trace of its presence in the vitreous. I do not
believe we should remove the eye at once in this case,
but make an iridectomy by removing that portion
of the iris that is wounded. If we find the foreign
body in that part, then we may feel well assured
that the patient will have nothing more to suffer
from than an ordinary iridectomy, unless there have
been other serious complications.

It is well known that a foreign body can remain
in the lens substance for an indefinite time, without the least apprehension of its presence causing

inflammatory action. This is accounted for by the absence of blood vessels and nerves in that structure. The lens will of course become opaque, but by retaining the injured eye, it may, at some future time,—in the event of loss of vision in the other eye, —luckily be a happy consequence to remove the lens in order to give the patient vision.

A foreign body may remain in the vitreous for a lifetime, causing no inflammation, and very little defect of vision; as a rule, however, inflammation and abscess of the vitreous—causing pan-ophthalmitis—are the results, with great danger of sympathetic ophthalmia and loss of the other eye.

It has been conceded on all hands, if the foreign body has lodged within the retina or the choroid, or has severely wounded the ciliary body, there is no alternative; enucleation should be done at once.

I desire to report the following interesting case which demonstrates the length of time a foreign body may remain in the eye without exciting much inflammatory action, either in the injured eye, or through sympathy with its fellow:

On June 9, 1890, I was consulted by Mr. T., who was suffering pain in and about the right eye, especially in the supra-orbital region. He also suffered from severe neuralgic pains in the right side of his head. His suffering had been of recent date.

His history of the case is as follows: He is a machinist by trade. About four years ago, while engaged in chipping a piece of iron, he received a

shock to the right eye, as if struck by the fist of some
person. So positive was he that he had received
such a blow, that he accused some of his fellow work-
men who were with him, of having committed the
offense. He rubbed his eye, and feeling that he was
not wounded or injured in any way, continued his
work, but discovered shortly that he had double
vision, and in consequence of this confusion of images, severely wounded his finger by a misdirected
stroke from his hammer, and was for this reason
compelled to lay off for a few days. This occurred
in a few minutes after the injury to the eye. About
two days after this event the eye became greatly inflamed and he suffered severe pain from it. The inflammatory action continued about four weeks,
when it gradually subsided. He now noticed that
vision in the eye was very imperfect, that there
were constantly floating before it six or seven spots
that were fixed, or, rather, in a relative position with
each other. Aside from this he suffered no inconvenience with the eye for the space of about four
years.

At the time the patient consulted me, there was
very little congestion about the eye. The iris, however, presented a somewhat greenish appearance, as
having undergone at some previous time a severe inflammation. There was a small corneal opacity to
the right of the pupil and immediately opposite a
small opening in the iris, indicating that these
structures had been transfixed previously with some

foreign body. The pupil was rather small and did not dilate under atropine.

With the ophthalmoscope the vitreous was quite opaque and filled with floating debris. On account of the smallness of the pupil and the opacity of the vitreous, it was impossible for me to discover any fixed foreign body, or the course it had taken upon its entrance into the eye.

As previously related, he saw six or seven fixed spots before his eye, but now he saw quite a number —about forty, as he counted them—that they were constantly increasing in numbers and that they were not fixed as formerly, but floated promiscuously among themselves.

With this history of the case, in connection with the examination of the injured eye, I concluded a foreign body had entered it, and advised him to have it removed. He objected to such procedure, and argued that as there was some vision in the eye, it should not be removed.

I detailed the danger to which he was subjecting himself of losing the other eye through sympathetic inflammation. He asked that I should prescribe an anodyne to relieve his pain, and went his way.

On June 15th he again returned to my office. There had been no cessation of the pain, but, on the contrary, he said that it was more severe and that vision in the other eye was not as good as formerly. He thoroughly realized his situation, and asked that the eye be removed at once.

On the next morning I removed the eye, which was found in the condition here presented. A chip of steel almost three-fourths of an inch in length was found protruding through the sclerotic, within less than one-fourth inch of the optic nerve entrance, about one-fourth inch of which showed on the inner surface and almost one-half inch on the outer, which had been buried in the cellular tissue behind the eye.

FIGURE 27.

This foreign body, after entering the eye, during the inflammatory stage that soon followed, had become encapsulated and thus for a long time remained there without causing further inflammatory action. But after a few years, perhaps on account of the oxidation of the iron, this encapsulated matter became disorganized and diffused throughout the vitreous humor. This condition is doubtless what caused the appearance to him of so many spots in his field of vision.

It is difficult to say what would have been the result had the eye not been removed. It is probable that a pan-ophthalmitis would have ensued, and that there would have been a suppuration of the humors of the eye.

Whether the other eye would have been affected through sympathetic inflammation is a question, but certainly the danger was too great to sacrifice it by delay in removing the injured one, in anticipation of its becoming a useful organ. But here is a lesson. The significance of a foreign body within the eye is a matter of serious consequences, not only to the eye which has directly received the injury, but to the other, which often suffers indirectly, because of sympathetic inflammation.

What is most misleading is the fact that the eye receiving the foreign body may in a great many instances retain some vision for a considerable length of time, but, as a rule, the sluggish processes of inflammation subtly, but surely, present themselves.

Observation has abundantly demonstrated the fact that neither eye is safe when one has a foreign body within it. But as long as the eye is not painful and has a respectable appearance, whether there remains in it some vision or not, many are inspired with a hope that vision may at some future time return or at least improve, and thus nurse a delusion which soothes them into tranquility, until finally they awaken to the fact that they are blind.

In case of severe injury to an eye, where the vision is entirely destroyed, and the ball is mutilated to such an extent as for its appearance to embarrass the patient, I am sure enucleation is preferable to its retention, in the hope of its making a suitable cushion for an artificial eye.

It is a great mistake to retain the stump of an eye, even if its muscles are not impaired, for the purpose of giving motion to an artificial eye, for there are many cases of sympathetic ophthalmia which have thus been incited.

From this consideration of the subject, we are enabled to conclude that an eye should be removed at once:

1. If it is severely wounded, so that we are positive under the most favorable circumstances it will be permanently disabled, whether there is or is not a foreign body within it. If the patient suffers pain, there can be no question as to enucleation.

2. When a foreign body has entered the eye and passed through the vitreous, there is no reason for delaying the procedure. Of course if it lies in the vitreous, it may become encapsulated and not be the cause of future trouble; but if it has lodged in the retina or choroid, or even passed through into the sclerotic, it is sure to be a source of mischief.

I might add that the fact of the foreign body being encapsulated does not insure immunity from future trouble, but the fact that a foreign body is within the eye, even if encapsulated, admonishes us to look with suspicion upon any pain about either eye and especially so if there is dimination of vision in the uninjured eye.

3. When the eye has a foreign body within it, no matter for how long a time, if there are constant neuralgic pains within or around either eye, even if vision in the injured organ is not entirely lost.

It occasionally occurs that, owing to an injury of the eye, the crystalline lens becomes dislocated and floats around against the iris and ciliary body, often causing considerable irritation. If this condition is allowed to continue, there is danger of sympathetic inflammation, to prevent which the lens should be promptly removed.

We should be fully impressed with this fact, that enucleation, if resorted to in time, will prevent sympathetic ophthalmia, but if delayed until the inflammation has commenced, is of little use.

ENUCLEATION OF THE EYE.

(Bonnett's Method as described by Meyer.)

"The patient having been placed on a couch and some anæsthetic administered, the surgeon takes hold of a fold of conjunctiva near the cornea, above the insertion of the internal rectus muscle, and incises it with curved scissors; then introducing the scissors beneath the conjunctiva, he frees the subjacent cellular tissue very completely. He next slips a strabismus hook beneath the muscular insertion, and cuts the tendon at a short distance from the sclerotic. Having done so, he continues to divide the conjunctiva, keeping near the cornea, till the next muscle is reached, which is detached from the sclerotic in a similar way, and so he proceeds till all the recti muscles are cut.

With a pair of strong forceps he then takes hold

of the eyeball in the sclerotic by the tendinous extremity of the internal or external rectus muscle,
which has purposely been cut somewhat long, and
drawing the eye as far as possible forwards and to
the side, introduces a pair of scissors, kept shut,
along the eyeball till the optic nerve is reached,
which he divides by a single cut.   (Fig. 28.)

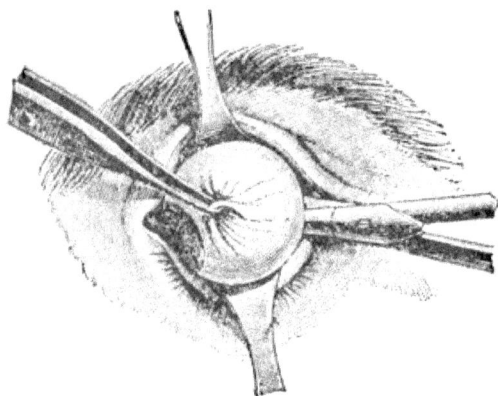

FIGURE 28. (After Meyer.)

If the operator be standing behind the patient,
he may prefer to divide the optic nerve of the right
eye from the temporal side of the orbit, and that of
the left from the nasal side.

As soon as the optic nerve is divided, it is very
easy to dislocate the eyeball and completely detach
the oblique muscles.   Some surgeons draw the edges
of the conjunctival wound together with a suture
drawn circularly through it and tied like the strings
of a purse.

The hemorrhage in this operation is insignificant; we apply an antiseptic dressing (lotion of sublimate, iodoform in powder dusted over the wound), and a pressure bandage. Cicatrization is generally completed in a few days."

ARTIFICIAL EYES.

After the wound from the removal of an eye has thoroughly healed, an artificial eye of the proper shape and size should be inserted.

Should the cavity remain for a considerable length of time after the operation without an eye, the muscles of the lid contract to such an extent that it is impossible to wear an eye of the size that will correspond to its fellow.

Care should be taken that the eye, when fitted, is not too large, in fact, the first eye should be somewhat smaller than the natural eye, for the reason that the muscles of the lid can be stretched to such an extent that the artificial eye hereafter fitted must be larger than the natural one in order to retain its position in the orbit.

The great advantage of an artificial eye beside its cosmetic effect is, that it places the lids in a proper position for the lachrymal puncta to receive the secretions of the eye, otherwise the tears would overflow and run down the face.

If there had been evidences of sympathetic inflammation or irritation in the sound eye before the removal of the injured one, we should be very careful about fitting an artificial eye, too soon, as it alone may excite a sympathetic inflammation, especially if there is a latent tendency to that condition.

Artificial eyes should always be removed on going to bed, and thoroughly washed and dried.

To introduce an artificial eye.—Before introducing an artificial eye, it should be placed in water; the upper lid should be elevated with one hand, while the artificial eye is held in the other hand and is inserted under the upper lid; the lower lid should now be brought forward and downward so as to admit the lower border of the eye to the cavity.

To remove an artificial eye.—It can be easily removed by inserting the head of a pin beneath its lower edge and bringing it forward.

CHAPTER XVII.

SECTION I.

ELEMENTARY OPTICS.

OPTICS is the science which treats of the nature and properties of light.

Light is the agent or force by the action of which objects are made visible.

Luminous bodies are those which emit light of their own generating.

Illuminated bodies are those that merely diffuse the light that they receive from other bodies.

Bodies are *transparent, translucent* and *opaque.*

Transparent bodies allow objects to be seen distinctly through them.

Translucent bodies admit light, but do not allow objects to be seen distinctly through them.

Opaque bodies cut off the light entirely and do not allow objects to be seen.

The science of optics is divided into *dioptrics* and *catoptrics.*

Dioptrics is that part of the science of optics which treats of refracted light.

Catoptrics is that part of optics which treats of reflected light.

In refraction the ray of light passes through the body, but is changed in its course.

In reflection the ray is not allowed to pass through, but is thrown off from the body on which it falls.

The word *refraction* is derived from the Latin *refrango: re* "again" or "back," and *frango, fractum* to "break," and signifies the deviation or bending of the rays of light from their original direction, in passing from one medium to another.

A *medium* is anything through which rays of light can pass, as the atmosphere, water, or the glass of which lenses are made. The *media* (plural of medium) of the eye are the *cornea, aqueous humor, crystalline lens,* and the *vitreous humor.*

A *lens* is a glass, which owing to its peculiar form, causes the rays of light to converge to a focus or disperses them according to the laws of refraction.

As we have to deal considerably with a convex lens within the eye, namely, the *crystalline lens,* some points in reference to convex lenses in general will be opportune.

The strength of a convex lens is governed by the distance that parallel rays of light (that is, rays of light from infinite distance) in passing through it are brought to a focus upon the opposite side (Fig. 29).

For instance, the rays of light in passing through this lens are represented as coming to a focus at 6 inches. The lens would therefore be called a 6 inch lens. Should the rays focus at 3

inches, a 3 inch lens, and so on; the sooner it would
focus, the stronger the lens.

The rays of light entering the lens are called
incident rays; and those emerging are called the
emergent rays.

Now there is a truth common to convex lenses,
that if the principal focus is doubled, the rays of
light coming from a point at that distance and passing through the lens will come to a focus at the same
distance on the opposite side. Thus we will take
a 6 inch lens and place an object at twelve inches
from its surface, then the rays of light emanating
from this object and passing through the lens will
come to a focus at twelve inches on its opposite side
(Fig. 30).

In the first instance where the parallel rays pass

FIGURE 29.   FIGURE 30.

through the lens and focus upon its opposite side, the
focus is called the *principal focus*. Any other focus
is called the *conjugate focus*.

With this knowledge of convex lenses, I will
now give you a rule that will enable you to determine
the point (conjugate focus) at which rays of light will
focus in passing through a lens, should they come

from any point between infinite distance (20 feet or more) and its principal focus.

From the preceding it is plainly seen that the rays of light which emanate from an object at its principal focus, on emerging from the lens will be parallel (Fig. 31). It will also be seen, if the object is placed at a point nearer than the principal focus, the rays in passing through the lens will become divergent as seen in Fig. 32.

This represents the only condition in which rays of light passing through a convex lens become divergent.

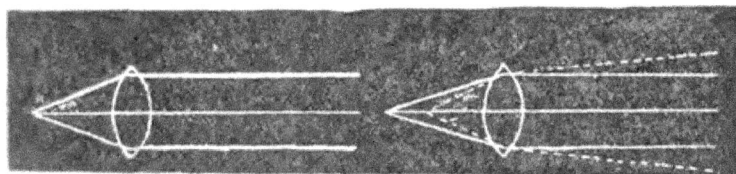

FIGURE 31.   FIGURE 32.

But rays of light from an object at any point between infinite distance and the principal focus will converge somewhere (the conjugate focus), and the following rule will indicate that point:

Multiply the length of the principal focus (with the proper signs annexed) by the length of the focus of the incident rays, and divide the product by the difference between the principal focus and the focus of the incident rays. The quotient will be the conjugate focus.

Example: Where will the rays of light focus in passing through a 6 inch lens, the object being placed at 9 inches from surface of the lens?

Answer: 9 x 6 = 54 ÷ 9 — 6 = 18 inches.

Concave glasses cause parallel rays of light in passing through them to become divergent (Fig. 33).

The focus of a concave lens is found by continuing backwards its emergent rays to a point where they all meet (Fig. 34).

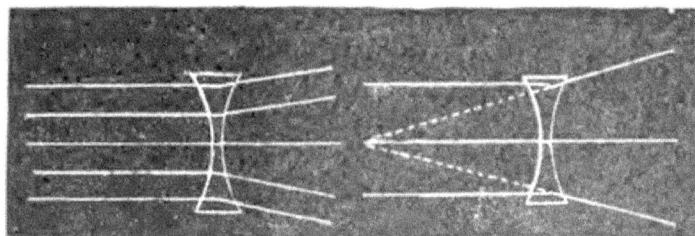

FIGURE 33.   FIGURE 31.

With this knowledge of lenses, I think we can properly appreciate the subject of errors of refraction.

The *errors of refraction* consist mainly in the length of the eye. The eye is either *too long*, or *too short*. Now, there is within the normal eye the crystalline lens, and the ciliary muscle that controls the shape of the lens in such a manner that when we look at near objects (10 to 18 inches distance) the lens becomes more convex, and when we look at objects at infinite distance (20 feet or more) the lens returns to its usual shape, and it is less convex than when looking at a near point. This is the natural conditon of the lens in the normal eye, and we see without any effort, that is, without "strain." This property of the eye in adapting itself to see at different distances is known as *accommodation*.

It follows then, the *nearer* the object is to the normal eye, the *greater* the accommodation or "strain;" the *farther* the object is away from the eye, the *less* the effort to accommodate or "strain."

Now, if rays of light coming from infinite distance, in passing through the crystalline lens, should focus upon the retina so as to form there a distinct image, then the eye is considered normal in its refraction or *emmetropic* (Fig. 35).

But, should these rays of light (parallel rays) focus in *front of the retina*, the eye then is *too long* and we would be compelled to use a concave lens

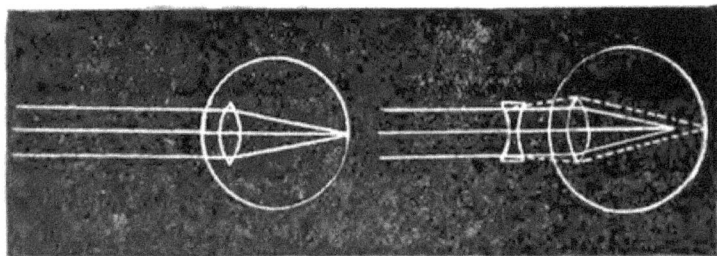

FIGURE 35.   FIGURE 36.

in front of the eye, that is, a lens that would disperse the rays of light sufficiently to cause them to focus upon the retina, and thus correct the defect (Fig. 36).

This condition of the eye is known as *myopia* or *short-sightedness*, for a reason which we will presently explain.

If, upon the other hand, these parallel rays of light should focus *behind the retina*, then the eye is *too short*, and we would be compelled to place a convex lens in front of the eye, so that the rays of

light may be converged sufficiently to allow them to
be focused upon the retina (Fig. 37).

This condition of the eye is known as *hyperopia*.

Now, if the crystalline lens of the normal eye
were fixed, and were not possessed of considerable
elasticity, it would be impossible for it (the normal
eye) to see objects plainly at a near point; objects
then would be seen distinctly at a far point only,
and the nearer they would approach the eye the less
distinct they would become.   But the eye is happily
supplied with the means to overcome this obstacle,
for the lens is so controlled that it is made to become very convex when the object is brought to a
near point, as has hitherto been explained (Fig. 38).

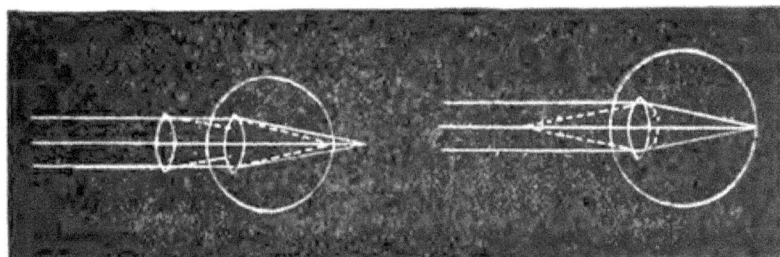

FIGURE 37.   FIGURE 38.

(The white lines represent the normal eye at rest,
that is, receiving impressions from a distance, and
the dotted lines from a near point.)   Thus the lens,
by becoming more convex than when receiving rays
of light from a distance, is enabled to focus those
from a near point upon the retina.

But the crystalline lens, at about the age of 45 years, commences to lose its elasticity, and continues to become less and less flexible as age advances, until 70 years is reached, at which time it is so hard that the accommodation is entirely lost, and no effort can exert any influence upon its shape.

This condition of the eye is known as *presbyopia* or far-sightedness, because the person thus affected can usually see well at a distance, while vision is very much interfered with for near work. This condition requires the addition of convex lenses for near vision, the strength of the lens being gradually increased as the years advance.

The property of a convex lens of thus increasing the conjugate focus of the emergent rays as the object is placed nearer to it, plays an important part in the errors of refraction, at one instance to the disadvantage, as in hyperopia, and at another to the advantage, as for near vision in myopia.

Let us examine into the mechanism of *hyperopia*, or that condition in which the eye is represented as being *too short*.

We have represented here parallel rays passing through the eye and coming to a point on the other side of the retina. How can these rays be made to focus upon the retina? We must either accommodate sufficiently to make the lens convex enough to cause the rays to meet at the proper point, or in the event the error is so great, or the ciliary muscle that

controls the lens is so weak that we cannot,—then
we must use in addition a convex lens (Fig. 39).

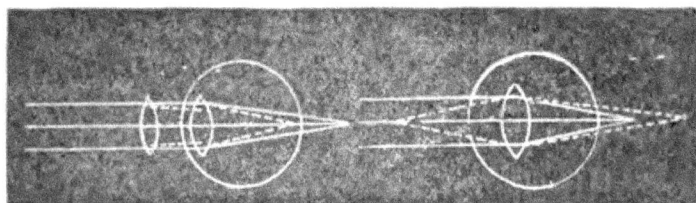

FIGURE 39.   FIGURE 40.

If we are compelled to accommodate in order
to see at a distance (focus parallel rays), then it follows that we will be compelled to accommodate much
more in order to see near objects, for we have been
taught that the nearer the object approaches a convex lens, the farther upon the other side will the rays
focus (Fig. 40).

It is not unusual for the patient, who frequently
has a considerable degree of hyperopia, to be able to
accommodate enough to cause the lens to become
sufficiently convex, that the rays of light may focus
upon the retina, for not only a distance, but for near
vision; but when this "strain" is kept up for a considerable time, the eye becomes greatly fatigued and
may suffer serious consequences.

There is a condition of hyperopia which is frequently mistaken for myopia. It is this: the hyperopic eye, as has just been explained, is compelled to
accommodate excessively in order to do near work,
and constant near work with such an eye causes the

ciliary muscles which control the lens to become
so cramped, that the lens is almost constantly in a
fixed condition, and that for near work; when the
lens is thus rendered very convex, the patient is not
in a condition to see at a distance, for the rays of
light in this case focus before reaching the retina.
(Fig. 41.)

In this case, as in myopia, a concave glass will
cause the rays of light from a distance to focus upon
the retina.   But this glass will not remedy the defect but for the instant, for the moment that there
is the least relaxation in the spasm of the ciliary
muscle, which is liable to occur at any time, then
the glasses do no good, but work a positive injury.
This condition of the eye I have called *factitious*
or *forced myopia*, and it is very common, especially
in overworked school children, and others who are
compelled to do excessive eye work.

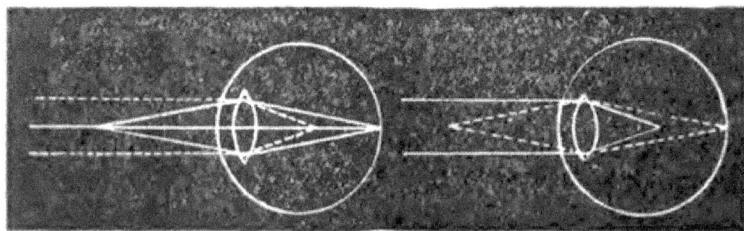

FIGURE 41.   FIGURE 42.

It now remains to explain how increasing the
length of its conjugate focus will be advantageous
for near vision in myopia.

The eye being too long in myopia, the rays of

light from a distance in passing through its lens become focussed before reaching the retina.

But we have been taught the nearer the object approaches a convex lens, the farther upon its opposite side will the rays of light which pass through it become united, as is represented in Fig. 42. (The white line represents the myopic eye, viewing objects from a distance, and the dotted lines at near work).

It is thus seen that in high degrees of myopia the patient is unable to see very near objects without an effort of accommodation; although, in fact, the least accommodative effort, in many cases, would cause a blurring of the vision.

To summarize: It has been demonstrated that the normal eye in a state of rest is adjusted for parallel rays of light, and that these rays focus upon the retina *without* accommodative effort; at a near point *with* accommodative effort. If the eye is too short, the effort of accommodation, for near work especially, becomes too fatiguing, and the muscle which controls the lens, like any other muscle of the body which is overworked, may become spasmodically cramped, thus inducing a latent optical defect; that is, there may be an error in the refractive apparatus of the eye, but it is so disguised by reason of the cramp, that it is impossible to detect it without a thorough examination of the organ during complete relaxation of its accommodation under atropine or other mydriatic. Otherwise, it is impossible to detect this error.

If the eye is too long, no action of the ciliary muscle can make the lens concave or adjust it so that the rays of light passing through it will focus upon the retina. To correct this defect, as we have hitherto explained, requires a concave glass.

The normal eye then receives impressions upon its retina from infinite distance, without strain or accommodative effort; from a near point with only ordinary strain or accommodative effort, or with such an amount of effort only as was designed in nature to accomplish this function.

It follows, then, if the media are clear and the retina normal, and the ciliary muscle is not disabled, that images will be received upon the retina and their impression carried to the brain with as little effort and as unconsciously as that of breathing, or as the action of the heart in the healthy individual.

There is another defect in the refractive condition of the eye, which, beside the defect in vision, is the cause of much nervous disturbance. This defect is known as astigmatism.

Astigmatism is caused by an irregularity in the curvature of the cornea, or crystalline lens.

Astigmatism is that condition of the eye in which the refraction varies in the different meridians.

In this condition of the eye, the rays which enter it along one meridian are brought to a focus before those which enter it along another meridian, thus the curvatures of its different meridians are not equal.

In astigmatism the meridian of greatest curvature is usually at right angles to those of the least curvature.

Astigmatism, then, may be defined as the inability to observe lines of the same intensity in one direction, as well as those at right angles to them.

When it is possible for the eye to observe the horizontal lines distinctly, and the vertical lines are blurred, but can be made distinct with a convex lens, then we have *simple hyperopic astigmatism with the rule*, for it is usual for a convex, instead of a concave glass to make the vertical plain.

If, however, a concave glass is required to make the vertical lines distinct, then we have *simple myopic astigmatism*, or astigmatism *against the rule*.

Astigmatism is always *with the rule* when a convex glass corrects the vertical line, and a concave corrects the horizontal. It is always *against the rule*, when a concave will correct the vertical line, and a convex the horizontal.

In *compound astigmatism* the eye is hyperopic or myopic in all meridians, but more so in some than in others.

Mixed astigmatism is that condition in which the eye is hyperopic in some meridian and myopic in others.

So far, our discourse has been directed to monocular vision, or to that of one eye alone.

We must also take into consideration that in order to have perfect binocular vision, both organs

must be alike; that is, they must have the same refractive powers; but if on account of one or more
of the muscles in one or both eyes becoming deranged, so that it causes one of the eyes to turn in,
or out, or up, or down, or in any manner away from
its normal visional line, the images then will not
properly merge, but will be more or less confused,
and the continual effort or strain required to accomplish the merging of the images is the fruitful source
of the headaches of which we hear so much, and
which are in no wise exaggerated.

Dr. Stevens of New York City, to whom should
be accorded the credit of making a practical classi-
fication of the different forms of muscular defects
of the eye, gives the following:

Orthophoria denotes parallelism of the visual
lines, or normal power of the muscles.

Heterophoria, non-parallelism of visual lines.

Esophoria, a convergence of the visual lines, or
insufficiency of the abductors.

Exophoria indicates divergence of the visual line
or insufficiency of the adductors.

Hyperphoria indicates the visual line of one eye
above its fellow.   *Cataphoria* below its fellow.

Hyperesophoria signifies a tending upward and
inward.

Hyperexophoria a tending upward and outward.

SECTION II.

THE CAUSES AND EFFECTS OF INSUFFICIENCIES OF THE OCULAR MUSCLES.

The normal functions of the orbital muscles, when the eye is considered as a monocular organ only, are complicated; when we study their actions in connection with binocular vision, their offices appear confused in the extreme; but when, in addition, there exists an abnormal condition in the action of one or more of these muscles, then we have a perplexing skein to untangle. Such has the subject of insufficiencies of the orbital muscles proven itself to be.

In seeking a remedy for an affection, we do so more intelligently by first searching for the underlying condition of which the symptoms are but the declaration. Hence in asthenopia of the orbital muscles, as in all other affections, we would, as far as possible, trace all symptoms back along the line of causation to their ultimate origin. We are thus necessitated to consider some of the most noticeable and common affections of the ocular muscles, whose conditions are attributed to refractive abnormalities.

For the purpose of being clearly comprehensible, the subject will be discussed in the following order:

1. Does an error of refraction contribute in any manner to muscular asthenopia?

2.   What is the *modus operandi* of the impairment of the function of the ocular muscles in ametropia?

3.   Will rendering the eye emmetropic contribute in restoring the weakened muscle to its normal condition?

The only difference between muscular asthenopia and strabismus is that in insufficiency there is temporary inability to maintain binocular vision, while in strabismus the inability is constant.   Muscular asthenopia implies an inability to bring both visual lines to bear constantly upon one point.   In strabismus there is inability to bring both visual lines to bear upon one point at any time.   In muscular insufficiency, then the muscle is partially disabled, and is enabled only a part of the time, and then with considerable effort, to perform its functions; while in strabismus it is totally disabled from performing these functions.

We are fully cognizant of the influence of hyperopia and myopia upon the induction and maintenance of convergent and divergent squint.

Observant oculists have noted that from 75 to 85 per cent of all cases of convergent squint are hyperopic, and in the divergent there is even a larger per cent of myopia.   This alone adequately demonstrates the influence of the ametropiæ upon the functions of the orbital muscles.

As we all well know, it does not follow that all cases of hyperopia and myopia are the subjects of muscular asthenopia, the occupations of the ametropic having much to do in developing this affection.

In order that the eye may deviate from its normal position, one of two conditions is necessary: there must be a physical or functional weakness of one muscle, or set of muscles, from which the eye is deflected, or an excessive strength of a muscle or set of muscles, toward which it becomes directed.

In simple hyperopia and myopia how is this accomplished?

The Modus Operandi.—In hyperopia the patient is compelled to accommodate in order to focus the rays of light upon the retina and make the image more distinct. The greater the degree of hyperopia the more he is compelled to exert his accommadation. Accommodation produces convergence, and the long continued effort of accommodation for this reason makes the convergence permanent.

In a few cases of hyperopia the muscle is not able to stand the long continued strain at convergence, and in order to avoid confusion of images or *diplopia*, the eye is instinctively turned out and entirely away from its fellow. Thus in hyperopia we occasionally have a divergent squint.

In myopia, as is well known, the patient sees well and often without effort of accommodation, when the object is brought to a very near point; hence, in order to obtain binocular vision, one or both eyes must become abnormally converged. The greater the degree of myopia, the more convergence is necessary in order to maintain binocular vision. The effort being irksome, and the work too fatiguing to accomplish with both eyes simultaneously, as is especially the case when there is much near work to perform, one eye is disregarded and involuntarily turns out, far enough away from its fellow, that there may be no confusion of images.

In those cases of myopia which are accompanied by convergent squint, the myopia is usually of small degree, and the amount of convergence is not so great but that the internal recti muscles may become orthopædically trained and strengthened by use for near work. In this case it is usual for one eye to become permanently convergent, and able to perform only near work, while it entirely disregards objects for a far point.

As myopia is the most common cause of divergent squint, it is evidently the greatest factor in the cause of insufficiencies of the internal recti muscles.

Myopia is *real* or *factitious*. It is *real* if, when there is complete relaxation of the accommodation, the measurement shows that the retina is situated behind the principal focus. It is *factitious* or forced myopia, if the ciliary muscle is cramped in such a manner as to cause the crystalline lens to become so convex that parallel rays of light in passing through, meet before reaching the retina. This condition, as we all know, is frequently acquired in eyes that are emmetropic, and even in small degrees of hyperopia.

So far we have had under consideration the causes whereby the muscles have become permanently deranged, so that we may be more thoroughly equipped for the discussion of that other condition of the ocular muscles wherein they become temporarily unable to perform their functions.

Now it is a matter of very little importance whether the eye is really or factitiously myopic, or is hyperopic, the conditions for the production of weak internal recti muscles are exactly the same; that is, in myopia, to recapitulate, the patient must place the object close to the eye in order to obtain binocular vision. The greater the degree of myopia, the more convergence, and the more constant the effort at convergence the more strain on the internal recti muscles; hence the asthenopia.

In hyperopia the patient accommodates to cause the lens to become sufficiently convex to focus the rays of light upon the retina. As accommodation causes convergence, the more he accommodates the more he converges, and the greater the strain upon the internal recti muscles.

We can converge without accommodating, but cannot accommodate without converging. If the patient could accommodate without converging, then there would be no strain upon these muscles, hence no asthenopia in hyperopia.

As heretofore mentioned, the internal recti muscles are most frequently affected with the inability to properly perform their functions; although the

external recti muscles are occasionally thus affected.
This condition is produced in one of two ways: either,
first, by a permanent contraction and increased
strength of the internal rectus orthopædically obtained—that is, acquired by moderate and continued
effort at convergence; or second, by a spasm or
cramp of the internal rectus caused by an over-ex-
ertion of that muscle.   In either case the result is
the same; the external rectus is weakened by the
long continued strain upon it.

I am now confident that this cramp or spasm
from overworked muscles plays a very important
part in the production of insufficiencies, and I am
also sure, because I have seen it practically demonstrated, that spasm of the ciliary muscle is an equal,
or perhaps the most important factor in the development of muscular asthenopia.

A muscle will not become weak without a cause.
Even if it becomes cramped and spasmodically contracted (functionally strengthened), it is an evidence
of weakness and not of tone.   Cramp or spasm is a
result of its weakened and overworked condition;
and for this reason I believe there are very few cases
of insufficiency without an error of refraction, either
real or factitious; and the forced error is certainly
the more productive of this condition.

Let us inquire how this may be accomplished.

Martin, I believe, advanced a theory of segment-
ary or unsymmetrical or, in other words, irregular
contractions of the ciliary muscles, whereby the lens

became irregularly curved or astigmatic.   We have good reason for believing that the theory is correct.

We have all of us seen eyes which have presented all the evidences of astigmatism, simple, compound, mixed, or irregular, which, after the accommodation had been thoroughly suspended under atropine, have been found to be emmetropic.   Now there was a functional defect somewhere before the accommodation was suspended, and that defect could have been in but one place—the crystalline lens.

Moreover, we occasionally meet with cases that do not show any evidence of astigmatism until after the accommodation is suspended.   In these there is corneal astigmatism which has been compensated for in the lens by its assuming such a shape as to correct the corneal irregularity.   Donders was the first, I believe, who brought to notice this condition.

We are taught that the macula lutea receives the impressions of images, and that it is the "sensitive point," that the images received at other points of the retina are not so distinct, hence it is the effort of the orbital muscles to so balance the eye that the impressions may be received on this particular point.

We know the influence upon the ocular muscles where there is a clear spot of cornea on an extensive opacity.   The eye involuntarily assumes that position in which the rays of light will best be received on this particular portion of the cornea, so as to fall as nearly as possible upon the yellow spot and thus secure the best possible vision that can be obtained under the circumstances. So also in other cases, where there is from any cause a removal of the pupil from behind the center of the cornea to some other position, the eye adapts itself to that position in which it can receive the best vision. To do this one muscle or set of muscles becomes strengthened, and their antagonists correspondingly weakened.

Now we know that any abnormal change in the curvature of the crystalline lens will change the angle of vision and, upon the same principle as in corneal opacities and abnormal positions of the pupils, disturb the equilibrium of the eye by its influence upon the ocular muscles.

If, then, certain changes in the curvature of the crystalline lens are productive of insufficiencies of the internal and external recti muscles, certain other changes will also account for insufficiencies of the superior and inferior recti.

To summarize thus: When an eye turns in, or out, or up, or down, or in any manner away from its normal position, there is one of two conditions existing. It does so either to place itself in such a position as to receive a better retinal image, and thereby assist both eyes to bear upon the same point; or to place itself in that position in which binocular vision will be entirely disregarded in order to avoid the confusion of images.

My experience has taught me that the low degrees of ametropia are, perhaps, more fruitful in the causation and maintenance of muscular asthenopia than the high, because of the strained effort to maintain binocular vision; for as heretofore explained, in the high degrees one eye is disregarded and the patient uses the other, while in the low degrees binocular vision can be maintained with some effort, which, however, if too steadily persisted in, produces fatigue.

From this consideration of the subject, the conditions which contribute to insufficiencies may be summarized as follows: Myopia, hyperopia, astigmatism and overwork, the latter being the exciting cause, which is more or less augmented by additional refractive error.

In considering the conditions which tend to the abnormal change of the visual angle, it would seem that the *punctum saliens* in the treatment of insufficiencies of the internal recti, is to prevent great and long continued efforts at convergence, and thus relieve the inordinate contractions of these muscles. Nothing will contribute so much to this purpose as remedying any refractive error, and thus rendering the eye as nearly emmetropic as possible.

The indications for treatment must necessarily be brief.

Although not entirely ignoring operative procedures, it must be acknowledged that tenotomies have not afforded the brilliant results so ardently claimed for them by their supporters. A tenotomy is prejudicial for one very serious reason: the movement of the eye is curtailed, and if the operation is

successful in effecting fusion of images at a near
point, there is usually too much restriction of motion
for a distance, hence an annoying diplopia follows,
which for persons engaged in the ordinary pursuits
of life is much worse than the insufficiency.

I am seriously impressed that operative procedures should be a *dernier ressort*, when all other
means have failed—then advancement of the weak
muscle, instead of tenotomy of the strong. There
is not then the risk of diminishing the movements
of the eye.

My experience has induced me to believe that
there is a very intimate relation between spasm of
the ciliary muscle and insufficiencies of the orbital
muscles. I rarely now make an examination for refractive error, without making the test for insufficiency, and I have been surprised at the frequency
with which I have met this affection.

Although insufficiency of the internal rectus is
far the most frequent, yet the test will frequently
disclose it in the other orbital muscles, and especially
in the superior and inferior recti.

The examination now referred to is rather a preliminary inspection, for we can rarely determine correctly the refractive condition of an eye without its
thorough atropinization. The accommodation requires to be thoroughly suspended before an attempt
is made at correction with glasses. This is important, whether there appears to be an insufficiency or
not in connection with the refractive error, for

the reason that after the eyes are thoroughly atro-pinized we usually find them in a very different con-dition from that shown at the preliminary examina-tion. What was then regarded as myopia, now turns out to be emmetropia, or perhaps hyperopia. What appeared a well-marked astigmatism is now neutralized, and insufficiencies frequently disap-pear.

When we consider the close connection which exists between the eye and the brain, we cannot wonder at the numberless forms of nervous sensa-tions and irritations which may result even from a slight ocular defect. The act of seeing is a very com-plex performance, requiring and exacting for its ac-complishment the harmonious coordination of a number of cerebral nerve centers; the second, third, fourth, and sixth, and the sympathetic nerves, in-dividually and collectively, take a part in the per-formance of this function, the least defect in any one of which will prevent that concordant unity of action designed by nature.

STRABISMUS.

The causes for insufficiencies of the external muscles of the eye have been fully discussed under the heading of the "Ametropiæ and their Relation to Insufficiencies of the Recti Muscles."

The operation for strabismus, when it becomes necessary to make it, is performed as follows: A fold

of the conjunctiva, near the margin of the cornea,
and over the insertion of the tendon of the muscle
to be divided, is siezed by a small pair of toothed
forceps, and is snipped by a small pair of blunt
pointed scissors.   Immediately, the capsule of Tenon

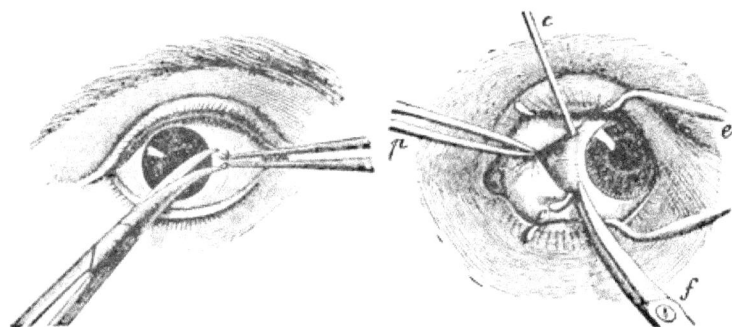

FIGURE 43.   (After Meyer.)

which lies under the loose conjunctival tissue, is
also siezed with the forceps and snipped with the
scissors in the same manner as was the conjunctiva.
The strabismus hook is now introduced and passed
under the tendon.   The conjunctiva over the point
of the hook should be gently pushed back with the
scissors, or some other convenient instrument, until
its point is brought through the wound.   We have
now within the curve of the hook the tendon, which
is divided with one snip of the scissors.   After this
has been done, the hook should again be introduced
in order to ascertain if the tendon has been completely divided.   If a few fibres should remain un-
divided, the operation may prove, on this account,
very disappointing.

It is unnecessary to unite the conjunctiva with sutures.

The only treatment necessary after this operation is to bathe the eye frequently with iced water that has previously been sterilized.

It is unnecessary to place the patient under the influence of ether previous to the operation except in case the subject is very young. When the ether is not used, the eye should be thoroughly cocainized previous to the operation by instilling into it a few drops of a four per cent solution of cocaine every two minutes for the period of a quarter of an hour.

SECTION III.

METHODS OF DETECTING AND CORRECTING REFRACTIVE ERRORS.

Notwithstanding the perfection which has been attained in a knowledge of the irregularities in the refraction and accommodation of the eye, yet a greater interest should be had by the general practitioner in a subject which has so much to do in augmenting the functions of one of the most important organs of the special senses.

There certainly seems to be a want of concern among physicians generally of those affections attributable to defects that are capable of correction with the aid of glasses, and too often they are considered of such insignificant importance as to be referred for correction to the jeweler, who makes some pretensions as an optician, or to the spectacle vender.

The eye presents many difficulties of study, especially that department which relates to departures from its normal refraction, owing to the necessary knowledge of optical principles, without which it is impossible to comprehend the abnormalities to which it is liable.

It is not expected of, nor is it possible for the general practitioner to know all about optics, nor is it necessary for him to know more than is taught of refraction and accommodation in standard works on physiology, to measure the length of the eye and estimate the irregularity in the curvature of the cornea, and in many cases adapt suitable glasses to these conditions.

A normal eye possesses in a wonderful degree the power of adjusting itself to view objects at different distances, but when it becomes embarrassed, as is too frequently the case, by certain abnormalities in its anatomical structure, by being too long, or too short, or in having an irregularity in the curvature of the cornea (which in the main constitute the errors of refraction), it is limited in its functions to such a degree that it is exposed to continual strains upon which depend so many of those distressing symptoms too frequently attributed to obscure nervous affections.

In order to appreciate the cause of a great many of these anomalies in its refractive power, we should impress ourselves with the property of the eye to adapt itself to different distances—in other words, its accommodation—regulated by the action of the ciliary muscle, which controls the lens in such a manner as to make it more or less convex, according as we desire to view objects near or far. In old age, of course, this action is limited, as the lens becomes less flexible.

It is the abnormal action of this muscle that plays such an important part in rendering indefinite and mystifying the refractive condition of the eye.

In order to equip ourselves as thoroughly as possible for detecting, as well as correcting refractive errors, we should remember, that the normal eye may, by reason of a spasm of the ciliary muscle, become seemingly abnormal in its refractive powers.

The above is one of the most important things with which we should impress ourselves on this subject, for it is through the ignorance or disregard of the above fact that so many people are suffering from the use of glasses that are not adapted to their eyes.

Defective vision, due to irregularities in the refraction of the eye, may be summarized as follows:

1. Defective vision may be due to structural imperfection, as irregular curved surfaces, or too short or too great length of the eyeball.

2. To a loss of the accommodation power of the eye.

The structural defects in the refraction of the eye are hyperopia, myopia and astigmatism.

Hyperopia is that condition of the eye in which it is too short.

Myopia is that condition of the eye in which it is too long.

Astigmatism is that condition of the eye in which the cornea or the lens, or both, are irregularly curved.

Hyperopia and astigmatism, or myopia and astigmatism, may be present in the same eye.

Loss of accommodation is due to loss of elasticity of the crystalline lens, or to a paralysis of the ciliary muscle.

Presbyopia is that condition in which there is a hardening of the crystalline lens, due to age.

General Considerations for the Fitting of Glasses.

In determining the refractive condition of the eye, it is necessary to have an outfit of appliances, more or less complete for this purpose, consisting of a trial case, test types, astigmatic cards, etc.

The trial case consists of a number of concave and convex lenses, either mounted or unmounted, single or in pairs, more or less complete as to the number of glasses employed, their variety in strength, etc.

A very complete case for the general practitioner is the Student's Trial Case, manufactured by

Meyrowitz, New York City. It consists of the following: 26 pairs convex spherical lenses, from .25 D. to 20. D. (2 to 144 inches); 26 pairs concave spherical lenses, from .25 D. to 20. D. (2 to 144 inches); 1 plane glass; 1 opaque glass; 1 red glass; 1 green glass; 1 single-grooved trial frame.

FIGURE 41.—Trial Case.

The physician desiring a more complete case will find all that can be desired in the Standard Trial Case, manufactured by the same firm, which contains the following: 30 pairs convex spherical lenses, from .25 to 20. D.; 30 pairs concave spherical lenses, from

.25 to 20. D.; 18 pairs convex cylindrical lenses, from .25 to 6. D.; 18 pairs concave cylindrical lenses, from .25 to 6. D.; 12 prisms, $\frac{1}{2}$ to 20 degrees; 4 plain colored glasses; 1 white glass; 1 half-ground glass; 2 metal discs, with stenopaic slit; 1 stenopaic disc, with hole; 1 solid metal disc; 1 improved adjustable trial frame, with revolving cells and graduated scales; 1 single-grooved frame.

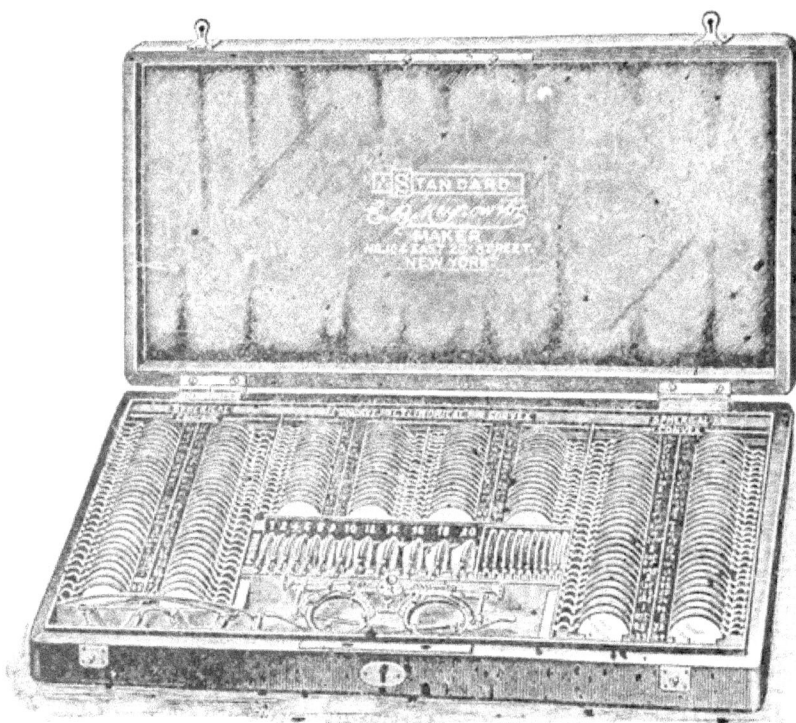

FIGURE 45 A. - Trial Case.

For the purpose of simply determining whether an error in the refraction of the eye exists, a very limited number of concave and convex lenses will suffice.

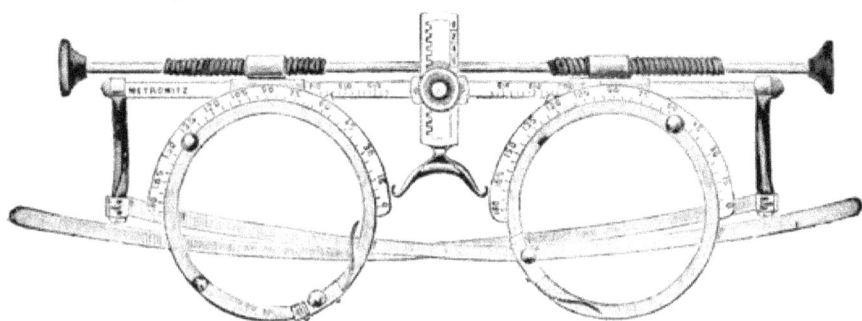

FIGURE 15 B.—Trial Frame.

As the testing of the eyes with a view of ascertaining a proper glass for a refractive error is a matter of great importance, we should not only be supplied with all of the necessary implements for determining any condition of refractive error, but we should have a well-lighted room, of proper length and sufficiently retired, that the physician, as well as the patient will have nothing to attract his attenion except that at hand.

The business of refracting eyes is such that no surgeon can perform this service with his office filled with lookers-on, or even with one or two who persist in annoying him in asking the patient questions while he is making the examination. They are usually friends of the patient, and continually persevere in "helping" you along, by saying "why that's a T," or "you called an O, C, or G," or "you can see better without glasses."

I have often been so nearly distracted by these

"helpful" people that I have learned to rid myself, in advance, of their presence by having them occupy another apartment, if possible.

A Snellen's card must be placed at a distance of 15 to 20 feet from the patient, however a test can be made at a shorter distance, if the room is not sufficiently long. The card should be placed directly opposite the patient, not facing him obliquely, the top of the card being not more than 5 feet from the floor, so that when the patient is seated on his chair the gaze, directed horizontally, will come into contact with the card at about its centre.

The same thing must be observed with regard to the astigmatic card. It must be placed, as nearly as possible, opposite the patient's eyes when he is seated and gazing horizontally, which will usually bring the test at about 3½ feet from the floor.

Z
S A
P X E
K F O L
R.T B V Z
L P H S N 4
O F Z D O E 3

FIGURE 46.
Test Type.

The advantage of having the card directly in front of the eyes, in the manner described, is readily observed if you will look at an astigmatic card obliquely; the horizontal lines, to an emmetropic eye, will appear very plain, while the vertical lines are correspondingly dull.

In Snellen's test card you will observe over each row of letters figures, "200," "160," "120," etc., down to "10." These figures indicate the distance at which the normal eye should be able to read the letters; for instance, over the top letter, or letters, you will observe the figures, "200." This indicates that the letters, over which these figures are placed, should be seen by the normal eye, unaided by glasses, at a distance of 200 feet. Consequently, if we have our patient situated at a certain distance from the card, and he is able to read the letters indicating the number of feet at which he is placed from them, then we infer that vision is normal.

The best astigmatic card is the "clock dial," as shown in the accompanying figure:

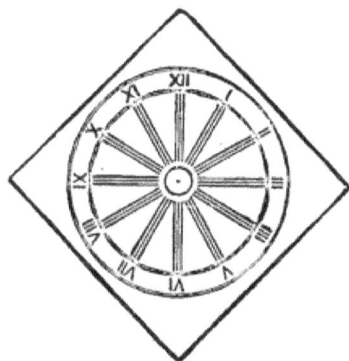

FIGURE 17.—Astigmatic Test.

We seat the patient opposite to, and facing, the test cards, and ourselves close by his side, and facing him. We also have the trial case conveniently near, on a stand.

If it is possible, as before mentioned, to place the patient at a distance of 20 feet, or even 15 feet, from the card, it is desirable to do so; but if this can not be done, then we can accomplish our purpose with a shorter distance, as 12 feet, or even 10 feet, for that matter.

We now place the trial frame, in one side of
which is placed an opaque disk, on the patient's face.
He is now asked to read the letters upon the card,
beginning at the top. He, at present, is supposed
to have the use of his accommodative faculties, that
is, his eyes are not under the influence of a mydriatic.

He may be able to read every letter on the card,
as far as the normal line, and still there may be a
serious error in the refraction of the eye. Such a
condition is often present, for the eye may make
itself appear to be normal, when in fact it is not,
through a strain, which condition will be explained
hereafter.

We will proceed farther. We now take a +.50
sph. from the trial case and place it in the frame,
and must notice attentively the effect. The patient
will probably say that he can still see the normal
line of letters, although not any better than with the
unaided eye.

We will now change the glass and put in its
place a still stronger one, say +.75 sph., and so continue until we obtain a lens which will cause the
letters to appear somewhat blurred and indistinct,
when we retrace and note the strongest glass, which
gives the best satisfaction, that is, that will make
the normal line appear distinct. This is called the
revealed or *manifest hyperopia*, and must be noted
in our record of examination for future reference.

We will change the disk to the other eye, and
proceed in the same manner as with the first one.

We will probably find that it will require the same lens to determine the amount of manifest hyperopia, as in the first eye. Fortunate if it does, and there is one thing of which we are assured: we can, with perfect safety, prescribe these glasses for the patient to wear, and if he is under 45 years of age, and has not lost his accommodative powers, the lenses will assist him very materially in his vision, especially for near work.

There can be no mistake in prescribing lenses in such a case as this. The strongest convex glass, that will make the normal line plain at its specified distance, will always assist vison for the near point, because it relieves the patient of an amount of strain equivalent to the strength of the glass, be it more or less.

Such an examination, as thus described, is termed the *preliminary examipation*, from the fact that every case indicating a defect in the visual powers of the eye, should have the accommodation relaxed with a mydriatic, and a very careful examination made while the eyes are under the influence of the drug.

If there has been serious trouble with the vision, it is positively indispensable to have the eyes placed under the influence of a mydriatic, which will thoroughly paralyze the ciliary muscles, especially if the patient is young or at any place under 45 years of age.

Of course, the older the patient the harder the crystalline lens and the less liable is it to be influenced by cramp or spasm of the ciliary muscle; but I have often been compelled to paralyze the accommodation in the eyes of patients over 45 years of age, to determine if there is not an astigmatism to correct, as well as a presbyopia. I have in such cases often been enabled to detect an astigmatism, which could not under other circumstances have been determined.

We now consider a case wherein the eyes have been atropinized, that is, the accommodation is paralyzed from the influence of a mydriatic.

We first have the patient read from the Snellen card, from a distance, without the aid of glasses, testing each eye separately.

We find by this examination that he can read distinctly as far as the line of letters over which is the figure 40, or the letters XL, indicating the number of feet at which the letters should be seen with the normal eye, unaided with glasses. We will say that he is seated at 15 feet from the card, and that he sees equally well with either eye. The vision in this case equals 15-40.

We now commence testing with glasses in order to ascertain if it is possible to improve his vision with glasses.

The trial frame is placed upon his face, with an opaque disk in it over the left eye. We now commence testing with a weak convex glass, say a plus 0.50 D. Upon placing this glass in the frame before the eye, we ask him to note the effect upon his vision.

He now, with the glass, distinguishes the line beneath, which should be seen, with the normal eye, at a distance of 30 feet, increasing the vision, so that it now equals 15-30.

As a plus glass caused an improvement, we continue, and apply next a plus 0 .75 D.   The vision is still improved, and is now made to equal 15-20; another glass is used, a plus 1 D, and vision is made to equal 15-15, or the normal, very readily.   We do not stop here, but continue to apply stronger glasses, and now apply a plus 1 .25 D.   Vision still remains at about the same point with this glass, but perhaps not quite so distinct.   We next apply the succeeding glass in strength, a plus 1 .50 D, and find that the vision is perceptibly blurred.

We now retrace, and by placing before the eye a plus 1 D, and then a plus 1 .25 D alternately, in tolerably rapid succession, we ascertain that the plus 1 D is the better, hence it is the strongest lens that gives the best satisfaction for distant vision. In the examination of the other eye, we find in all probability the same condition.

A question now confronts us: What is the strength of the lens that we shall prescribe for the patient?

One thing must be taken into consideration in all of these cases where a mydriatic is used, and that is, there is extreme relaxation of those muscles concerned in the act of accommodation, and for this loss of muscular tonicity an account must be taken, when we come to prescribe lenses for the patient.

There are a few oculists who do not take into consideration this abnormal relaxation of the ciliary muscle, and prescribe a glass, to what is called the *full correction*, that is, they prescribe the lens that gives the best satisfaction for distant vision, while the eye is under the influence of a mydriatic.

If at the preliminary examination, before the eyes were placed under the effects of the mydriatic, there had been revealed 1 D of hyperopia, then we can safely prescribe the full correction (plus 1 D); but, if on the other hand a weak concave glass improved the vision, then we can not safely prescribe the full correction, for it will probably over correct the hyperopia, and the patient will not see well at a distance with the glass when the effects of the mydriatic have fully passed off.

I have always felt that we should have some set rule to govern us under the above circumstances, and have followed this plan: If the eyes at the preliminary examination, that is, without the mydriatic, show the same degree of hyperopia as under the mydriatic, then I prescribe to the full correction; if, however, they do not show the same degree of hyperopia, then I deduct about one-third of the hyperopia shown under the mydriatic, and in the above case would prescribe a plus 0 .625 D.

If the examination, instead of revealing 1 dioptre of hyperopia, had revealed 1 .50 D, then I would have prescribed a plus 1 D; if it had been 3 D, then a plus 2 D should be prescribed; if 6 D, which is a very high degree of hyperopia, then I would prescribe a plus 4 D.

There is no one who can always prescribe a satisfactory lens, especially at one trial, no matter what rule is followed for selecting lenses, for while some eyes will accept the full correction, others can not, which makes it necessary that the patient should be under the observation of the oculist, often for a month or two, or until his eyes have become adapted to the new condition in which they are placed.

Another important matter in the selection of lenses may appropriately be taken up here; there is a condition of the eyes known as *anisometropia*, or that state where the eyes show a difference in their refractive powers; for instance, one eye will accept a plus 1 D, while the other accepts a plus 2 D.   Under these circumstances, what lens should be prescribed for each eye in order to secure the best vision and establish the best possible co-ordination of action between the two eyes?

My rule has been, under such circumstances as the above, to prescribe for the first eye the full correction, that is, a plus 1 D, and give a half correction for the other eye, that is, a plus 1 D, for each eye.

Fortunately it does not often occur that there is a very great difference in the refractive powers of the eyes in hyperopia, especially if they have been placed under the influence of a mydriatic; in fact it rarely occurs, although it is frequently observed before the mydriatic has been applied.   This is usually occasioned by a cramp or spasm of the ciliary muscle, and the same amount of cramp does not always exist in both eyes at the same time.

The matter of prescribing lenses of different strengths for each eye in hyperopia, is impracticable, for the simple fact that the glasses thus prescribed will usually cause a discrepancy in the size of the images, and the images thus formed will not merge, except under a strain which is as injurious to the patient as the hyperopia itself.

In *myopia* the same procedure in fitting for glasses is observed as in hyperopia, except that in testing without a mydriatic, we select the weakest concave glass that gives the most perfect vision. However, a concave glass should never be prescribed without placing the eyes under the influence of a mydriatic, for the reason that the crystalline lens may, on account of the eye having been taxed for a long time for the performance of near work, become so fixedly convex, that myopia may be simulated, when in fact the eye may be ametropic, or even hyperopic.

There is another reason for relaxing the accommodation in myopia, and that is, a latent astigmatism may exist in connection with the myopia, which could not be properly determined otherwise, and which, if corrected, would often add very much to the visual acuity.

It will be remembered, that in hyperopia it is not practicable to prescribe lenses of different strengths for each eye.   In myopia, however, this is frequently accomplished to a very great satisfaction, and the vision in both eyes materially benefitted.

The reason for this is twofold: In myopia the accommodation has not been used, or if so, to a very limited extent, while in hyperopia it has been practiced, often to an unusual degree.   In myopia, after the correction, the eye must assume the function of accommodating for near work, if it has been fully, or nearly so, corrected, and it does so gradually, as the ciliary muscle gathers strength, while in hyperopia the tendency is to retain its old habit of cramping, and thus over corrects, and that unsymmetrically, that is, in each eye in an unequal degree.

As the eye in myopia, especially of high degree, is more or less injured at its posterior section on account of the sclerotic giving way at this point and thus interfering with the normal contour of the retina in its most impressionable locality, the vision can not, as a rule, be brought up to the acuity as in a high degree of hyperopia.

After the patient has been fitted for distant vision, he often finds it very difficult for his eyes to become adapted to near vision, for the reason before mentioned, that the ciliary muscle, not having been in use, has lost, or never has had the requisite tonicity to induce the lens to become sufficiently convex to permit the rays of light from a near point to focus upon the retina.   It is for this reason that we must prescribe the weakest concave glass that will secure the best distant vision.

Where the patient has always been myopic in a high degree, the accommodation can never be entirely recovered, especially if the patient is pretty well matured before a correction has been attempted; for this reason the eyes should be refracted at as early a date as possible, and as myopia is a progressive affection, because of the constant lengthening of the ball, especially in the young, the vision should be frequently tested, and the glasses changed whenever the condition demands it.

Astigmatism being caused by an irregularity in the refracting surfaces of the cornea or the **crystalline lens**, is much more difficult to correct than **either** hyperopia or myopia, from the fact that we may have both conditions to deal with in the same eye.

As in hyperopia, we are liable to have spasm of the accommodating muscle, there is the same danger of over-correction, therefore the eyes should, in suspected astigmatism, be thoroughly placed under the influence of a mydriatic, and the accommodation completely relaxed, before an attempt is made to adjust lenses.

The effect of astigmatism, when it is present, is that the rays of light which pass in at one meridian, focus at a point different from those that enter at another meridian, the greatest difference being in opposite meridians. For instance, the rays of light may so enter the eye that the horizontal lines on the astigmatic card will appear plain, while the vertical lines are somewhat blurred.

As a rule, we will find in astigmatism, when the lines in any one direction, no matter what that direction may be, are plain, those in the opposite direction are usually blurred to a greater or less extent.

It is my desire to make this subject as attainable as possible, and to accomplish this end we will take up a case in the different forms of astigmatism, and thereby endeavor to detail the manner of correction with the trial lenses.

The eyes having been placed under the influence of a mydriatic, the patient is seated, as before, at fifteen feet from the test card. The clock dial card I consider the best for all forms of astigmatism.

In making the examination, especially if an astigmatism exists, it is very important to have a suitable trial frame in which to place the lenses. The necessity of this can be appreciated when it is understood that in some cases of astigmatism the change of a cylindrical glass to the amount of a degree or two, out of its proper place, often causes very much derangement in the visual acuity.

Figure 45 B is well suited for this purpose. See page 293.

The trial frame being adjusted on the patient's face, and the left eye covered with a disk, he is requested to look at the card, and asked to inform you which set of lines on the astigmatic card appears the plainest to him. He readily informs you that the horizontal lines are plain, while those directly opposite, that is the vertical lines, are blurred.

As in testing for any refractive error, we begin
with a weak convex glass, say a 0.50 D, by placing
it over the right eye. He is now asked to note the
effect which it has produced. He informs you that
the lines are changed, and that they are all somewhat blurred, but that the vertical lines are not so
much blurred as they were with the unaided eye.
We now remove the glass and place a convex 0.75 D
before the eye. He now informs you that the vertical lines are very plain, and that the horizontal lines
are very much blurred.

If we now remove the spherical lens and place
in its stead a convex cylinder of 0.75 D, axis vertical,
we will ascertain that the patient sees all of the lines
in the wheel equally plain, and that his visual acuity,
for a distance, is considerably increased.

In hyperopia, it will be remembered, you were
taught to make a deduction of one-third the strength
of the lens giving the best satisfaction under the influence of a mydriatic. In astigmatism, however, it
is different, and we fit to the full correction, and in
this case would prescribe a convex 0.75 D, axis 90
degrees.

It occasionally occurs that the axis of a cylinder
must be placed in some other position than the vertical, and it is a rule in hyperopic astigmatism, that
the axis of the convex cylinder is placed at 90 degrees, or within 5 degrees of that, but there are cases
where it is necessary to place the axis at 180 degrees,
the horizontal, or at a point anywhere between the
perpendicular and the horizontal.

Myopic astigmatism is corrected in the same manner as the hyperopic, but we use a concave glass to bring out the dull lines.

As has been indicated in myopic astigmatism, we will find that the axis of the correcting cylinder must, as a rule, be placed at 180 degrees, the horizontal, or within 5 degrees of that point; but, as in hyperopic astigmatism, it occasionally varies to a considerable extent, and it is occasionally necessary to place the axis at 90 degrees, or at a point anywhere between the vertical and the horizontal.

The character of the astigmatism so far noted is termed *simple astigmatism*. In simple astigmatism, with the unaided eye, the lines upon the clock dial appear clear and distinct in some one direction, and correspondingly dull in the opposite.

If the horizontal lines appear plain with the unaided eye, and the blurred lines are corrected with a convex cylindrical lens, the axis of which is placed vertically, then he has *simple hyperopic astigmatism*, and if the proper correction has been made, he will see all of the lines, running in the different directions, equally plain, and his vision will, for a distant point, be brought to the normal.

If, on the other hand, the vertical lines appear distinct with the unaided eye, and the horizontal lines are blurred, and if the indistinct lines can be corrected with a concave cylindrical lens, with its axis placed horizontally, then we have a case of *simple myopic astigmatism*.

If all the lines on the dial card are blurred, and it is necessary to use a convex spherical lens in connection with a convex cylinder in order to make perfect visual acuity, then the defect is known as *compound hyperopic astigmatism*.

How to correct *compound hyperopic astigmatism*.

We will take a case: The patient can not see any of the sets of lines upon the clock dial distinctly; we place before the eye a convex 0 .50 D; this glass makes all of the lines more distinct, but the horizontal ones more so than the others; we next apply a convex 0 .75 D, which makes the horizontal lines very plain, and the vertical lines plainer than before. We now find that the last glass makes the horizontal lines appear very plain; which make a note of, so that it is not forgotten. We continue our examination by placing the next strongest convex lens, a plus 1 D. We now find that the patient does not see the horizontal lines so plain as with the former lens, but the vertical lines appear plainer, yet they are still somewhat blurred; we remove this glass and place in the frame next a convex 1 .25 D, and the patient now informs you that the horizontal lines are very much blurred, but the vertical are very plain and distinct.

We make a note of the above examination, and we find that a convex 0 .75 D makes the horizontal lines very distinct, and that a convex 1 .25 D makes the vertical lines distinct, from which we deduce the following correction, viz: plus 0 .75 D, spher., with plus 0 .50 D., cyl., axis 90 degrees.

The above combination is now placed in the trial
frame, as indicated in the prescription, to verify the
inference, after which, in order to ascertain positively that we have not over-corrected or under-cor-
rected the eye, we place a weak spherical convex lens,
say a 0 .25 D, in front of the combination already
given, and the patient will readily inform you
whether it changes his vision or not, and in what
manner it does so.   If it aids his vision, we apply
another stronger glass, a plus 0 .50 D, which he will
probably say makes the vision less distinct.   Should
we in this manner discover that a change in the first
prescription is necessary, we make it accordingly.

Before making out our prescription, we must not
forget that our patient's eyes have been under the
influence of a mydriatic, and that a deduction from
the spherical lens must be made, of at least one-third
of its strength, which would make the spherical glass
a convex 0 .50;   then our prescription would be
changed to the following: plus 0 .50 D, spher., with
plus 0 .50 D, cyl., axis 90 degrees.

In all probability an examination of the other
eye will reveal the same refractive error, and will
require the same combination to make the correction.

In *compound myopic astigmatism* the same procedure may be resorted to in order to make the correction, by ascertaining the lens that will make the
lines distinct in one direction, then that one which
will render them distinct in the opposite direction,
from which a calculation can be made, and a lens

composed of a sphere and a cylinder prescribed,
which will correct both.

In compound myopic astigmatism, as in myopia,
we prescribe a glass to the full correction, basing our
selection upon the weakest lens that gives the best
results.

Mixed astigmatism is that condition of the eye in
which it is hyperopic in some of its meridians and
myopic in others.

Mixed astigmatism is corrected by a convex or
a concave spherical lens with a cylinder of opposite
power, as where a convex spherical lens is used to correct one meridian, a concave cylinder is used to correct the opposite meridian.

Mixed astigmatism is also occasionally corrected
with crossed cylinders of opposite powers.

Presbyopia is a loss of a certain portion of the
power of accommodation on account of the hardening of the crystalline lens due to age.

When the loss of the accommodative power has
reached to such a degree that near vision is difficult,
then it is necessary to use a convex lens of sufficient
strength to make objects distinct at a near point.

At about the age of forty-five years near vision
in the normal eye becomes somewhat indistinct, and
the patient adjusts his work at a more distant point
than formerly in order to see more distinctly.

As soon as presbyopia is disclosed, it should be
corrected in order to relive all possible strain for near
work. I am very sure that many cases of choroiditis

have been precipitated by straining the ciliary
muscles in order to see in presbyopia.

If there is an astigmatism existing with the presbyopia, it should be corrected in the same manner
as heretofore indicated in the correction of that defect; then with the correcting astigmatic lens in the
trial frame before the eye, a spherical lens can be se-
lected that will bring distinct vision to the desired
point.

It is impossible here to give the various methods
with a description of the many devices which are employed in correcting the many complicated defects
in the refraction of the eye. There are so many contingencies in the variety of forms of refractive error,
together with the accompaniment of spasm of the
accommodation, that nothing short of practice and
skill, backed by a thorough knowledge of optics, will
enable one to give satisfactory results in the fitting
of glasses.

TESTS FOR INSUFFICIENCIES OF THE OCULAR MUSCLES.

One of the oldest tests for the purpose of determining the presence of an insufficiency of one or more
of the ocular muscles consists in placing a lighted
candle at from twelve to fifteen feet from the patient.
Direct your patient to look attentively to he light
with both eyes, at the same time place before one
eye a red glass; if two objects are seen, a red and a

white one, it indicates that there is an insufficiency
of one or more of the ocular muscles.

There is one caution to be taken in making this
test, and that is, the patient must look squarely at
the light, not taking a side look, otherwise a satisfactory test can not be made in this manner.

From this test we can, by taking a little time,
determine any case of insufficiency of the ocular muscles, but it is a little tedious, and we must practice
more care than in making the test with some of the
improved instruments which have been invented for
this purpose.

In the above case, we say that two lights are
seen, a red light and a white one. Now, the positions of these lights will determine to a greater or
less extent which muscle or set of muscles is un-
balanced.

We say that the red glass is placed before the
right eye, and the patient notices two lights, a red
light and a white light, and that the red light is to
the right of the white one, but on the same line.
This position of the lights indicates that there is an
insufficiency of the external rectus muscle of the
right eye, or of the external rectus of the left eye, or
both, which we will determine hereafter.

We now place the red glass before the left eye,
in the same manner as we did before with the right
eye, and we ascertain that the red light stands to the
left of the white one, which verifies our diagnosis that
there is an insufficiency of one or both of the external
recti muscles.

How will we now proceed to make these two images merge?

If we will remember that to place a prism of proper strength, with its apex in the same direction in which the eye deviates, that it will cause the images to merge. With this knowledge we can very readily determine with a set of prisms which one will bring the images together.

We allow the patient to remain in the same position as when we made the last test, that is, the left eye is covered with the red glass, and in this condition he sees the red light to the left of the white light. In this connection one very important matter must be impressed upon your minds: in whatever direction the image is from the normal, the eye is directed to the opposite. For instance, as in this case, the left eye being covered with the red glass, the red light is shown to the left of the white light, the external rectus of the left eye being weak, it allows the left eye to turn in towards its fellow, but according to a well known law of optics the image is projected outwards.

With this information we now take a weak prism, say a 1 degree prism, and place it in the trial frame over the right eye, while the red glass still remains over the left one. As the image is projected outwards, the eye is directed in, therefore we place the apex of the lens in, and we ask the patient to inform us how the lights appear to him. He will probably inform you that they stand closer together, but they are yet separated. We now take a stronger prism, and by repeated trials ascertain the prism that will merge the images. When they are properly merged, there will appear but one light.

Should we conclude to prescribe for the patient, we divide the strength of the prism between the two eyes; if it requires a 3 degree prism to merge, then we would give a 1½ degree prism for each eye, recollecting that the apex of each must be directed in.

Insufficiencies of the other muscles, as the internal recti, or the superior and the inferior recti, may be determined in the same manner as those just considered.

It is occasionally necessary to correct two sets of muscles as an internal or external rectus, with a superior or an inferior rectus. In the above case, the image of one eye would be projected to the right or left and at the same time above or below that of the other.

There are other simple and convenient tests for determining insufficiencies of the ocular muscles, as the Maddox and the Stevens test, herein described:

The *Stevens test*, as described by himself, is as follows:

In the determination of the various tendencies of the ocular muscles it is often advisable, and even necessary, to bring to our aid as many forms of evidence as can be made subservient to our purpose.

While the phorometer remains pre-eminently the reliable and efficient working instrument in the determination of heterophoria, auxiliary means are often required to confirm or to explain its indications. We sometimes also require an instrument for making provisional examinations more portable than the phorometer.

As such an auxiliary and provisional instrument I have devised the stenopaic lens, which posesses manifest advantages (Fig. 48 A).

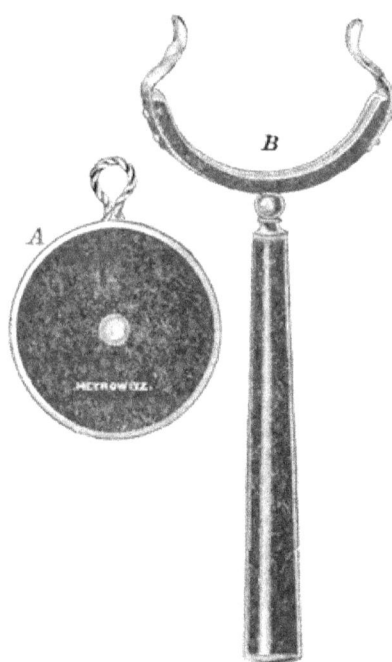

FIGURE 48.

The purpose is to present contrasting images to the two eyes.

With the lens the image of a candle flame twenty feet distant, seen through the stenopaic opening, is a large and perfectly defined disc of diffused light.

If, for the purpose of effecting a diffusion, we employ the uncovered convex lens, a very slight movement of the lens, in or out, up or down, gives to it the effect of a prism in those various directions.

If a convex lens, about 13 D., is covered, except at the optical center, where a circular opening of three millimeters or less diameter acts as a stenopaic window, the small opening serves the double purpose of preventing an adjustment of the lens as a prism and of cutting off the halo in such a manner as to give the impression of an exact disc of light bordered by a frame.   A metal or hard rubber disc of the size of the lens of the trial case, perforated by an opening of the required diameter and supplied with a perfectly centered lens, is a convenient form.   It may be used with a handle (Fig. 48, B), enabling the patient to hold it in his own hand, or it can be placed in the trial frame.

In orthophoria the untransformed image should be found exactly in the center of the disc.   In heterophoria it will tend toward or beyond the border. If the flame sinks below or rises above the center, while at the same time it deviates laterally, we thereby discover by a single comprehensive view all the elements of a compound deviating tendency, so far, at least, as that tendency is manifest (Fig. 49). In this important respect the stenopaic lens presents a feature both unique and of much significance. While by other methods of inducing diplopia or contrast we may discover, at a distance of some meters,

first one and then the other element of a deviating
tendency, by this instrument all the collective elements are presented simultaneously to the eye, thus
eliminating a very important source of error.

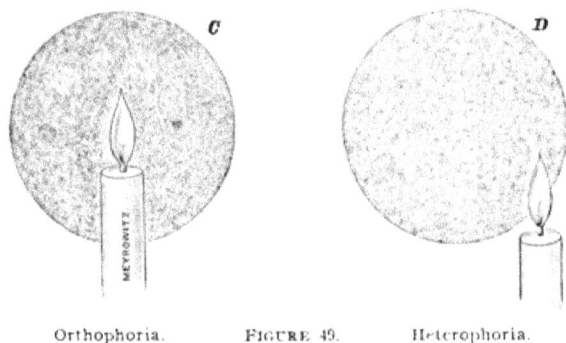

Orthophoria.   FIGURE 49.   Heterophoria.

In respect that it is simple, cheap and small
enough to be carried in the vest pocket, and that it,
more than any of its class, represents the true relation of the visual lines, it is a useful test. Its disadvantages are those common to every instrument held
close to the eye when in use in these examinations.

The *Maddox test*, as described by the author,
consists of a hard rubber disc mounted in a metal
rim of the size of trial lenses, so as to fit easily into
the trial frame, which holds in the center a glass rod.

The effect of this transparent cylinder is to cause
an apparent elongation of a single flame into a thin
line of light, quite dissimilar from the flame itself,
as seen at the same time with the other eye, so that
there remains practically no desire to unite the two
images, whose relative position thus easily indicates

the condition of equilibrium of the two eyes. The
line is always at right angles to the axis of the rod,
so that to produce a vertical line, with which to test

FIGURE 19 E.

horizontal deviations, the rod is placed horizontally,
and to produce a horizontal line, to test vertical deviations, it is placed vertically. The test is made
prettier and any desire for single vision still further
reduced by placing a red glass before the other eye.

CHAPTER XVIII.

COLOR-BLINDNESS.

THE following description of this subject is as given by Duane in his *Students' Medical Dictionary:*

"Color-Blindness. Blindness for one or more kinds of color. According to the Young-Helmholtz theory, it is due to the absence or failure to act of one or two of the three percipient elements in the retina, each of which is sensitive for one of the primary colors (red, green, violet), thus producing red-blindness (the most common form), green-blindness, or violet-blindness.

According to Hering, it is due to the absence from the retina of one or two of the three primary substances whose assimilation and disassimilation under the action of light give us the sensation of red and green, blue and yellow, black and white. If the red-green substance is absent, there is red-green blindness (= red - blindness — green - blindness of Young-Helmholtz); if both are absent, there is total color-blindness.

Color-blindness is tested for by presenting to the patient samples of different colors which are apt to be confused by the color-blind, and telling him to place together all that are alike."

CHAPTER XIX.

EXTERNAL EXAMINATION OF THE EYE.

BEFORE making an examination of the eyes, it is well for the physician to inquire as to the general health of the patient.

The cutaneous surface of the lids should be inspected first. Any inflammatory condition of the skin should be carefully noted, as the cuticle of the lid is liable to the same affections of the skin as that covering other portions of the body.

The edges of the lids should next be examined as to whether the cilia are properly placed. The position and condition of the puncta should be noted, together with the lachrymal sac. It is well to make pressure over the region of the lachrymal sac, so that any contents may be expelled through the puncta.

The conjunctival surface of the lower lid should be examined for granulations, foreign bodies or any signs of inflammation.

The upper lid should now be everted and its conjunctival surface inspected. To evert the upper lid,—the physician standing behind the patient who has been seated, directs him to lean his head back so that the top of the back of the chair acts as a head-rest. The patient is told to look down, and while so doing, the physician grasps between the thumb and index finger of the right hand the cilia situated at about the middle of the edge of the upper

lid. The index finger of the left hand is now placed
above the tarsal cartilage of the lid, the remaining
fingers of the left hand resting on the patient's forehead, and while making gentle downward pressure
with this finger, the lid is everted by an upper movement of the hand which grasps the cilia.

The lid having been everted, the condition of the
conjunctival surface should be noted as to the presence or absence of inflammation, granulation, foreign bodies, scars, etc.

The physician will often experience great trouble in everting the lids of children. In such cases a
few drops of cocaine should be instilled and the child
placed on the physician's lap in a recumbent position. A lid elevator should be used to avoid unnecessary pressure upon the eye. If this fails, it is necessary to etherize the patient.

The cornea should next be examined and its
transparency, the presence or absence of scars, ulcers, etc., carefully noted.

Oblique illumination is used in examining the
lens, anterior chamber, iris, and cornea. The patient is seated to the side and about three feet from
the light. The rays from the light are focused,
through a three-inch convex lens, upon the patient's
cornea. It will be seen that the rays, after passing
through the lens, are brought to a focus and form
a cone, the base of which is at the convex lens, and
the apex on the cornea. By moving the lens in different directions, the apex can be made to fall upon
any portion of the cornea. By the use of this method
the condition of the anterior chamber and any opac-

ities in the cornea and lens can be observed. Adhesions of the iris, if present, can be readily seen.

The iris and pupil should now be inspected. The sensitiveness of the iris to the influence of light and shade can best be observed by placing the hand over one eye; upon doing this, the pupil of the uncovered eye will be seen to dilate, if the eye is normal. On removing the hand from the eye, both pupils will be seen to contract to their former size.

To determine the intra-ocular tension, the physician directs the patient to close the eyes; he may then, by making pressure upon the ball with the index and middle fingers of either hand, determine the degree of hardness. Instruments have been devised to determine the intra-ocular tension, but by the continued use of the fingers the tension can be much better obtained, and with much less trouble.

The perception of colors is very important and is useful in determining the presence or absence of color-blindness. (This method is described more fully in another chapter.)

CHAPTER XX.

SECTION I.

THERAPEUTICS OF THE EYE.

MYDRIATICS.

A MYDRIATIC is a drug that causes dilatation of the pupil. The principal mydriatics are the alkaloids, *atropine, duboisine, hyoscine, gelsemine, coniine, cocaine,* and *homatropine.*

These mydriatics, with the exception of cocaine, produce relaxation of the ciliary muscle and paralyze the sphincter fibres of the iris. Cocaine acts by producing contraction of the blood vessels and by stimulating the sympathetic nerve.

The sulphate of atropine is probably the best mydriatic, and is used in the strength of two (2) to four (4) grains to the ounce of distilled water.

Homatropine. The following paragraph on homatropine is from *Hare's Therapeutics:*

"Hydrobromate of homatropine, properly applied by frequent instillations, is a reliable mydriatic for the correction of anomalies of refraction in healthy eyes. Experience is not at hand to determine its value for this purpose in eyes affected with retinal-choroidal disturbance. Atropine and hyoscyamine are preferred under such circumstances, for the obvious reason that their prolonged action is desirable as a method of treatment. The danger of systemic disturbance from homatropine is far removed, even when repeated instillations have been made, and its temporary action upon the pulse causes no inconvenience to the patient. Slight hyperæmia of the conjunctiva almost invariably follows its use, but true conjunctivitis, if it occurs at all, must be excessively rare. According to the studies of Dr. de Schweinitz and the writer, the drug has a physiological action closely allied to that of atropine, from which it is derived. Homatropine mydriasis generally lasts from thirty-six to forty-eight hours, that of hyoscamine eight to nine days, and that of atropine ten to twelve days. For the production of ordinary mydriasis, the drug should be used in solution of the strength of 4 grains to the ounce of distilled water, which is dropped into the eye every five or ten minutes. As the drug is expensive, only a few drachms of the solution of the strength named should be ordered for the patient."

Duboisine can be used in the same strength as atropine, but it is a more powerful drug than atropine and should be used with caution.

In case of poisoning from atropine, morphine is the antidote.

Mydriatics are contraindicated in all conditions of the eye associated with abnormal intra-ocular tension, as in glaucoma.

MYOTICS.

A myotic is a drug that causes contraction of the pupil. The most important myotics are *eserine*, *pilocarpine, physostigmine, morphine, muscarine,* and *nicotine.*

Myotics produce spasm of the ciliary muscle and as a result paralysis of the accommodation until the effects of the drug wear off.

Myotics are used in glaucoma to reduce the intra-ocular pressure, and are somtimes used after cataract operations to prevent prolapse of the iris.

LOCAL ANÆSTHETICS.

Cocaine hydrochlorate is used as an anaesthetic in the eye. It is usually used in a solution from one to four per cent.

A new anaesthetic, called *euacine,* has been placed upon the market, but it has not yet been thoroughly tested by the profession.

ASTRINGENTS.

The *glycerite of tannin* is used as an astringent. It is used in blepharitis marginalis and in other forms of irritation of the lids caused by the acrid secretions from the different forms of the ophthalmias. It should be applied with a brush or cotton swab.

Solutions of the nitrate of silver are used in varying strength. In granular lids, a solution of from one to two grains to the ounce can be employed. In cases of purulent opthalmia, in a solution of ten grains to the ounce, the nitrate of silver can be used in the adult, but I doubt its efficacy.

STIMULANTS AND ALTERATIVES.

The ointment of the *yellow oxide of mercury* (Unguentum Hydrargyri Oxidi Flava) is a very valuable preparation and is applied to the inner surfaces and edges of the lids. It is indispensable in blepharitis. Care should be taken that the preparation be properly mixed, for if the oxide has not been reduced, the ointment is useless and its application a positive injury to the patient's eyes.

It is used in the strength of from one to five grains of the yellow oxide to the ounce of vaseline.

A good eye wash for patients suffering with a simple conjunctivitis or conjunctival irritation from any cause is the following:

R.—carbolic acid, 1 g tt
        hydrastin, grs ss
        boracic acid, grs x
        cocaine hydrochlorate, grs iv
        glycerine, drs ii
        aqua hamamelis, drs vi
        Mix and filter. Sig.—A few drops in the
            eye three times a day.

ANTISEPTICS.

The *bi-chloride of mercury solution*, 1-5000, is probably the best antiseptic that can be used in eye affections. It should be used in all purulent ophthalmias. The best way to get its good effects is to flush the eye with the solution by means of a small syringe, care being taken that the syringe has a blunt nozzle, so as not to injure the eyeball. A few drops of cocaine should be introduced into the eye previous to the flushing.

Carbolic acid is a very useful remedy in eye affections. It is stimulating, antiseptic, and anæsthetic. It can be used in a solution of from one to three drops of the acid to the ounce of glycerine.

Boracic acid can be used as an antiseptic in the strength of five grains of the acid to the ounce of distilled water.

SECTION II.

THE USE AND ABUSE OF LOCAL MEDICATIONS IN EYE AFFECTIONS.

Relying upon the advances in bacteriology, the causes of many diseases which were formerly considered obscure, or were based upon theory alone, have been established, and the old materia medica of many affections has been modified or entirely replaced by that which science and experience have demonstrated as the most useful.

The science of bacteriology has benefited no branch of medicine more than that which relates to the diseases of the eye, for it has demonstrated that the most formidable inflammations to which the eye is subject are the products of bacterial or allied organisms.

What had been gained hitherto, in the way of mitigating and subduing these inflammations, was

effected more through empiricism than by the
rational method of a study of the causes, and the
consequent deduction therefrom of remedies.

When Crede, a surgeon of the Lying-in Asylum
of Leipsic, recommended that a ten-grain solution
of the nitrate of silver be instilled into the eyes of all
new-born children in order to prevent ophthalmia
neonatorum, he did not probably base its prophylactic properties upon the theory that it was a germicide; but it had become known that strong solutions
of the nitrate of silver would cure purulent ophthalmia, and upon this fact alone I presume Crede based
his theory of preventing the affection. The medical world rejoiced, however, that a remedy had been
discovered that would prevent that terrible affection which has caused so much misery and filled the
asylums with blind children.

An explanation of the manner in which the nitrate of silver solution prevented purulent ophthalmia was made by a prominent oculist after the gonococci in purulent conjunctivitis had been discovered, which was that the nitrate of silver solution
acted as a cautery to the superficial layers of the epithelium, and thus destroyed the gonococci which had
penetrated these layers, and which, he claimed, the
milder disinfectants could not reach.

Although Crede demonstrated that about 11 per
cent of cases sank to 0.1 per cent by the use of the
nitrate of silver solution, yet there are those who
have made an industrious study into the cause and
the treatment of purulent ophthalmias who question the advantage of the procedure, and in fact

doubt not that its breach would be better than its observance.

The skilled oculist of to-day recognizes that a ten-grain solution of nitrate of silver in the eye of an infant is dangerous, and will often set up an inflammatory action as destructive as ophthalmia neontorum itself. We must remember that the cornea of a new-born child is delicate and immature; its structure is not compact; the epithelium is easily abraded; and the sclerotic is so thin and loose in its structure that the reflex from the choroid is easily seen through it. Under these conditions even a five-grain solution is considered too strong, and it is now admitted that a two-grain solution properly applied is as effective a prophylactic as the stronger solutions, and that a 1-5000 solution of bichloride of mercury is far better than the silver solution, and much less liable to injure the child's eyes.

Under the old regime, the bluestone and solutions of varying strengths of nitrate of silver, sugar of lead and sulphate of zinc were indispensables. No one thought of making up a prescription for the eye without one or more of these as ingredients, no matter what the cause, whether due to the infection of a bacillus, or to the effects of an hereditary dyscrasia.

There is scarcely an affection of the eye in which these old remedies, the bluestone, the sulphate of zinc and sugar of lead are adaptable, but on the other hand are positively injurious. They should be expunged from the ophthalmic materia medica.

Two things have contributed to revolutionize ophthalmic therapeutics; one, the discovery of a local anesthetic applicable to conjunctival surfaces, and the other, the revelation that the purulent ophthalmias are the products of a specific bacteria, and the value of antiseptics in their treatment.

At the present time it is conceded that the very best treatment in the incipiency of a purulent ophthalmia, is to flush the conjunctival sac with a mild bichloride of mercury solution, and if this treatment is kept up several times a day by washing all particles of pus out of the sac, the discharge will cease in two or three days, and the conjunctivitis will subside in a remarkably short time.

If the ophthalmia is well under way when we have been called into the case, the treatment is the same, only it may be necessary to use the antiseptic solution oftener; the discharge will gradually cease, and the tendency is to recovery.

Simply instilling this solution into the eye from a dropper is not sufficient, for in this way it does not come into contact with but a part of the conjunctival sac, especially if there is much swelling of the lids and conjunctival chemosis. It should be applied with a small syringe, the nozzle of which has been flattened so as to be readily intruded between the lid and the ball, pressure being made toward the lid to prevent injuring the ball. Before flushing, a few drops of cocaine should be instilled into the eye.

Other antiseptic solutions may be admissible in the purulent ophthalmias, but I doubt if any known remedy has the combined antiseptic and germicidal properties of the bichloride solution.

Atropine is of no value in ophthalmia neonatorum or any other affection of the eye of an infant, even if there is a condition demanding a mydriatic, for the reason that it will not dilate the pupil of an infant.

While atropine is susceptible of doing great mischief if used in the eyes of those suffering from subacute or chronic glaucoma, yet there is no doubt that its use is too much restricted, especially by the general practitioner.

The symptoms of an incipient iritis are often so indefinite that it is impossible to differentiate them from those of a catarrhal conjunctivitis, until great injury has been accomplished. In every case of conjunctival irritation with a contracted and sluggish pupil, atropine should be instilled into the eye until there is thorough dilatation. This procedure would save many eyes. Excluding the purulent ophthalmias, iritis stands at the head of the list in the destruction of vision.

Carbolic acid is a great remedy in painful affections of the eye, especially in that resulting from corneal ulcer. It can be applied in from gtt.j to gtt.iij to the ounce solution. The carbolic acid solution serves a three-fold purpose: It is stimulating to the abraded cornea, it is antiseptic, and it is anesthetic. As an anesthetic, its effects are almost instantaneous, and are very lasting, and is specially adapted in cases of corneal ulcer of the phlyctenular variety.

SECTION III.

ANTISEPSIS.

In order to obain the most favorable results not only in operations upon the eye, but also in the treatment of the various affections of the eye, it is necessary that strict antiseptic precautions be resorted to.

Previous to an operation the hands of the surgeon should be thoroughly scrubbed and all dirt removed from under the nails. After thoroughly cleansing the hands, they should be placed, while yet wet, into a hot corrosive sublimate solution, 1-1000, for a short time. It is necessary that the assistants also take the same antiseptic precautions that the surgeon does.

The skin around the eye to be operated upon should be thoroughly cleansed, after which the sublimate solution should be aplied, care being taken that none enters the eye. After cocainizing the eye of the patient, it should be flushed with a bichloride solution, 1-5000. It is also advisable to place a towel wrung out of the bichloride solution 1-1000 around the patient's head, so as to cover the hair.

The preparation of the instruments is another important matter. Cutting instruments should never be boiled as the boiling damages the edges. It is well, however, to boil non-cutting instruments. The instruments should never be placed in a bichloride solution for the reason that it will rust them. Probably the best way to render the instruments aseptic is to cleanse them in alcohol and then place

them in a 1-20 solution of carbolic acid for about
ten minutes, after which they should be placed in a
bowl of hot water and they are ready for use.

BANDAGES.

The bandages and dressings should be clean and
aseptic, and care should be taken that they are not
so heavy as to produce undue warmth. They should
be removed when soiled and fresh ones applied.

To make a compress bandage, place a piece of
absorptive gauze over the closed eye and upon this
place enough cotton so that when a roller bandage
is wound around the head, there will be mild pressure on the eye.

GENERAL CONSIDERATIONS OF OPHTHALMIC THERAPEUTICS.

There are many minor details with regard to
the therapeutics of the eye which if considered alone,
would appear insignificant, but when taken collectively are of the utmost importance.

In all eye affections, it is well to warn the patient not to strain, but to rest the eyes.

If the light pains the eyes, coquells should be
ordered.

Protective spectacles should be used to protect
the eyes from the wind and dust when necessary.

Shades are valuable to shield the eyes from the
bright light.

Natural leeches are of no value in eye affections,
and should never be used, as they are liable to cause

septic infection. If there must be blood-letting, the artificial leech should be resorted to.

Applications of heat and cold are of great value in eye affections. Cold should be used only in recent wounds and in the very incipiency of inflammation. After the inflammation has become established, fomentations of hot water are to be used.

Poultices should never be used for the reason that they are septic.

Tea leaves, rotten apple, beef and all such articles that have been recommended by the laity are not only valueless, but are positively injurious to the patient because of their septic tendencies.

Patients suffering from any inflammatory affections of the eyes should have their own napkins and towels which should not be used miscellaneously.

No physician should allow a wash, the contents of which are unknown, to be applied to the eye.

Patients who are suffering from gonorrhoea should always be warned of the purulent character of the discharge and its dangerous effects, if allowed to come in contact with the eye.

Lead or zinc lotions should not be applied under any circumstances, as they are liable to form permanent opacities.

Caution should be used in employing atropine in old persons, as it sometimes checks the secretions and causes retention of the urine.

The physician should never put atropine in an eye where there is increased tension.

CHAPTER XXI.

RETINOSCOPY.

RETINOSCOPY (Shadow-test; Skiascopy; Fundus-reflex test; Keratoscopy;) is a method of determining the refraction of the eye by examining the movements of the shadow when the fundus oculi is illuminated by light thrown into the eye by a moving mirror.

The method of directing the rays of light by means of a mirror through the pupil to the fundus of the eye, and determining the refraction of the eye by the motions of the shadow following these rays, in conjunction with the movements of the mirror, is an old method resurrected and described under a new name.

This method was best described by MacNamara in 1882, under the name of keratoscopy, as follows:

"Keratoscopy is best explained by placing a convex lens at such a distance from a screen that rays of light passing through the lens from a concave mirror are focused on the screen. Under these circumstances a small and bright image of the lamp is formed, with a sharply defined and dense surrounding shadow. If the lens is now brought nearer to, or removed from the screen, the image of the lamp becomes feebler, and the line of demarcation between it and the surrounding shadow is fainter. As the lens is moved to different distances from the screen the variations between the brightness of the image and the surrounding shadow will be evident. At the same time, supposing the concave mirror from which the light is reflected is rotated, the direction of the image and shadow will be seen to move on the screen in a direction opposite to that in which the mirror is rotated. If this principle is applied to the human eye, and light reflected from the surface of a concave mirror is directed into the eye, on looking through the sight-hole of the instrument we percieve an illuminated area surrounded by a deep shade on the retina. But since the image is seen through the media of the eye, the direction in which the image moves as the mirror is rotated will depend upon the refraction of the eye under examination."

With this explanation of the principles involved, the method is readily demonstrated.

The mirror used may be plane or concave and has a small central sight-hole.

FIGURE 50.—Plane Mirror for Retinoscopy.

A shade is also necessary, which should fit over the lamp, and at a point opposite the flame is a hole through which the rays to be reflected by the mirror pass. The surgeon sits beside the light in such a position that the rays emerging from the aperture in the shade are reflected by the mirror into the patient's eye.

If the pupil is very small, it should be dilated with homatropine previous to the examination. After throwing the light into the patient's eye, by rotating the mirror in different directions, there will be seen movements of the light on the fundus oculi.

The light should be thrown into the eye at an angle of about 15 degrees.

If the mirror used is a plane one, the movement of the shadow will be found to be with the motion of the mirror in hypermetropia, emmetropia, and myopia of 1 D or less.

In myopia of more than 1 D, the shadow moves against the motion of the mirror.

If the mirror used is a concave one, the movement of the shadow will be found to be against the motion of the mirror in hypermetropia, emmetropia, and myopia of 1 D or less.

In myopia of more than 1 D, the shadow moves with the mirror.

The manner of examination is as follows: The trial frame is placed on the patient's face and the movement of the shadow is noted. (A concave mirror of about a twenty-inch focus being used.) If the eye is myopic, the shadow will move with the mirror. Now by placing a concave lens in the frame and by increasing its strength each time until the shadow moves against the motion of the mirror, the degree of myopia can be found.

Thus, if it requires a —4 D lens to make the shadow move against the motion of the mirror, the degree of myopia is 4 D.

To this must be added —1 D, because at this distance of four feet the observer strains and thus adds about +1 D to his own eye.

If it were possible for the surgeon to make the observation at about twenty feet, or to entirely relax his own accommodation, it would not be necessary to make this addition of —1 D.

As the shadow moves against the motion of the mirror in emmetropia, hypermetropia, and myopia of less than 1 D, it is necessary for us to determine which of these three conditions is present.

In emmetropia, the shadow is very distinct and the light area is clear and bright with well defined borders, while in hypermetropia there is a blurry, crescentic shadow with indistinct borders. In hypermetropia we begin with a +1 D and increase the lens until the shadow reverses its direction and moves with the mirror.

In hypermetropia +1 D must be subtracted from the lowest plus glass which reverses the direction of the shadow, for the defect is over-corrected.

If a +1 D will cause the shadow to move with the mirror and the direction of the shadow will be the same with a lens less than +1 D, then the eye is slightly myopic less than —1 D.

But if a +1 D causes the shadow to move with the mirror and a lens less than +1 D causes it to move against it, it is emmetropic.

If there is a difference in the motion of the shadow in two opposite meridians, it is indicative of astigmatism.

If it is simple or compound astigmatism, there need not be any great difficulty in determining the amount, but if it is mixed, it cannot be accomplished by this method.

The best manner for ascertaining the astigmatism is to correct each of the principal meridians separately with spherical lenses, after which you can calculate the compound lens necessary for the correction of both meridians.

This method of determining the refraction of the eye is not practical, especially in the intelligent adult, and should be resorted to with children only, or those who are not well enough informed to read letters and appreciate the intensity of lines at different angles.

CHAPTER XXII.

SECTION I.

OPHTHALMOSCOPY.

T is well known that all of the rays of light that enter the eye are not absorbed, but that some of them are reflected back out of the eye. These rays that are reflected back out of the eye pursue the same course as that by which they entered.

It will be readily seen that if the observer could place his eye in the direct path of these reflected rays without interfering with the source of light, a view of the fundus oculi could be obtained. This difficulty was overcome by Hemholtz in 1851 by the invention of the ophthalmoscope.

The ophthalmoscope, as invented by Hemholtz, consisted of several plates of glass held together by

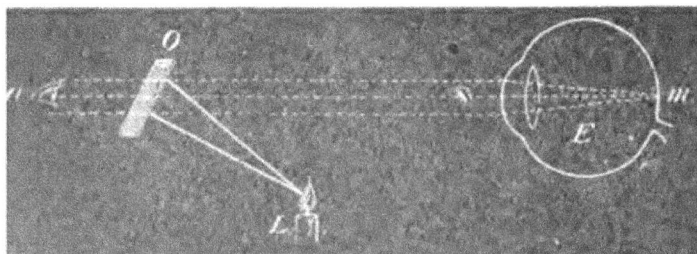

FIGURE 51.

a frame. The rays from the light L (see figure) are reflected into the patient's eye by the plates at O.

The rays being reflected back out of the eye from M, strike the plates at O, and some of them pass on through the plates to the observer's eye at *a*, and the point *m* is made visible to the observer.

Many improvements have been made on the ophthalmoscope since the days of Hemholtz, so that the modern instrument bears but little semblance to the one used in those days. Of those in use at the present day the Loring ophthalmoscope is one of the best and consists of a tilting oblong mirror with a small central aperture through which the observer looks. The ophthalmoscope is provided also with a circular disc in which are placed concave and convex lenses of varying strength. This disc is so placed that it can be revolved by the finger during the examination and thus brings before the observer's eye the desired lens.

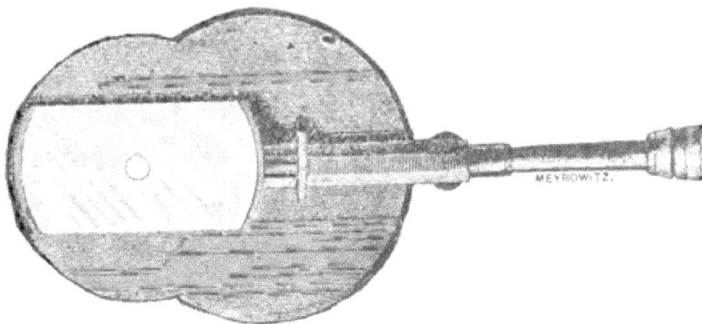

FIGURE 52A.—Loring's Ophthalmoscope.

The purpose of these lenses is to neutralize any error of refraction either in the observer or patient, or both, so that a good view of the fundus can be obtained.

Previous to an examination with the ophthalmoscope, it is necessary to dilate the pupil with some
mydriatic.   The light should be placed behind the
patient's head and on the same side as the eye to be
examined, care being taken that the patient's eyes
are in the shadow.   The observer sits facing the patient and gazing through the sight-hole of the ophthalmoscope directs the rays reflected from the mirror into the patient's eye.

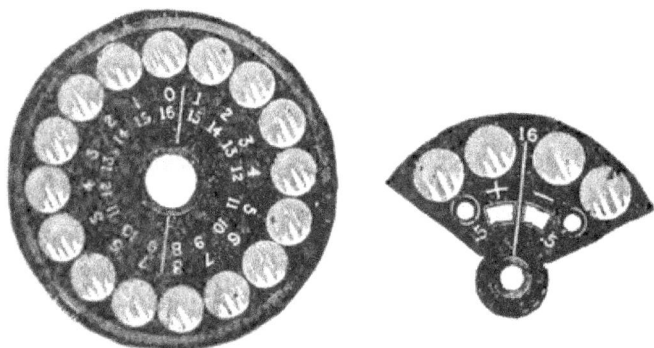

FIGURE 52 B.

There are two methods of examination with the
ophthalmoscope, the direct and the indirect.

THE DIRECT METHOD.

In this method of examination, the parts of the
fundus are seen erect and magnified about fourteen
or fifteen diameters.

The observer places his eye very near to that of
the patient, in order that a more extensive view of
the fundus can be obtained.   If a mydriatic has not

been used, the patient should be directed to look into space so as to relax his accommodation, and it is necessary for the observer to do likewise, for if either accommodates, the rays will be brought to a focus in front of the observer's retina, and in consequence a blurred image. This is one of the essential points in making an examination with the ophthalmoscope, and until the student can learn to relax his own accommodation, the results will be very unsatisfactory.

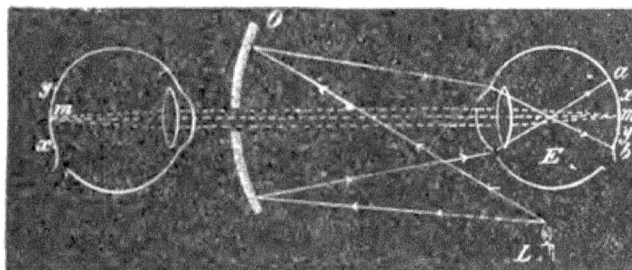

FIGURE 53.

On examining the figure, the formation of the image in this method can be readily apprehended.

The rays reflected from the concave mirror O converge, and, on passing through the refractive media of the eye, become more convergent and focus in the vitreous humor and illuminate the retina from *a* to *b*. While some of these rays are absorbed, others are reflected back out of the eye from all points included between *a* and *b*, and if the patient's eye is emmetropic, the rays from any point, *x*, for instance, will become parallel on emerging from the patient's eye, and on entering the observer's eye focus at *x*¹, providing his eye also is emmetropic.

In like manner rays from y become parallel and focus on the observer's retina at y^1, while those from m unite at m^1 in the observer's eye.

If the patient's eye is hyperopic, the rays from the point x diverge on emerging from the eye and the observer, in order to render these rays parallel, must place a convex glass before his own eye.

As the reflected rays are convergent in a myopic eye, the observer, in order to render these rays parallel, must place a concave glass before his own eye.

THE INDIRECT METHOD.

(Method of examining the inverted image.)

In this method a more extensive view of the fundus is obtained than by the direct method, although the different parts appear smaller.

The observer holds in one hand, between the thumb and index finger, a convex lens of about 16 D, close to the eye of the patient and steadies his hand by placing the remaining fingers on the brow of the patient. In the other hand is held the ophthalmoscope, about two feet distant from the patient's eye.

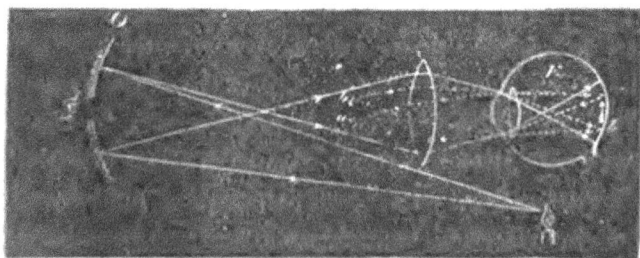

FIGURE 54.

On examining the figure this method can be readily understood.

It will be seen that the convex lens condenses the rays of light from the mirror and brings to a focus the reflected rays emerging from the eye, thus forming an inverted, aerial image of the fundus between the lens and the mirror.

There are many minor details with regard to the examination of the eye with the opthalmo-scope, which are omitted for the reason that it is impossible to intelligently explain them, as this knowledge can be obtained only by a continued and careful practice with the instrument.

SECTION II.

THE NORMAL FUNDUS OCULI AS SEEN WITH THE OPHTHALMOSCOPE.

Its Color. The color of the fundus oculi varies in different races. Among the white races it has a reddish color, while in the dark races it is of a grey-ish hue. It also varies in different members of the same race, for in very fair persons the fundus is very light, while in dark persons it is correspond-ingly dark. This difference in color is due to the amount of pigment deposited in the choroid; thus in the fair races there is usually a very scanty deposit of pigment, while in dark races it is very profuse. Owing to this profuse deposit of pigment in dark per-sons, the vascular structure of the choroid is with

difficulty discernible; on the other hand, in light
complexioned persons, owing to the small amount of
pigment, the choroidal vessels can be seen with ease.
There is a congenital absence of this pigment in some
people, which condition is known as *albinism*, and
persons thus affected suffer very much from the light.

Its Size. Owing to the magnifying power of the
refractive media of the eye, the fundus appears enlarged about fourteen or fifteen diameters.

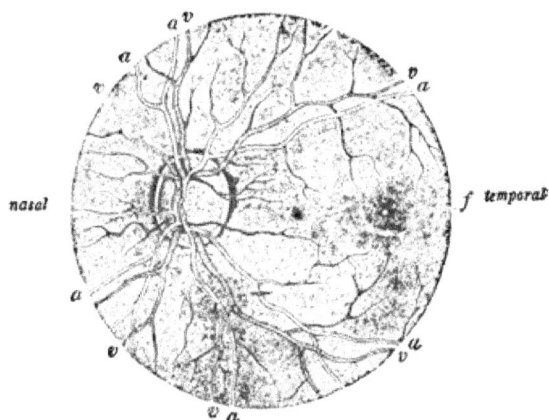

NORMAL FUNDUS OF THE LEFT EYE, SEEN IN THE ERECT IMAGE.

The optic disk, which is somewhat oval longitudinally, has the point of entrance of the
central vessels somewhat to the inner side of its center. That portion of the papilla
lying to the inner side of the point of entrance of the vessels is of darker hue than
the outer portion; the latter shows, directly to the outside of the vascular entrance,
a spot of lighter color, the physiological excavation with fine grayish stippling,
representing the lacunae of the lamina cribosa. The papilla is surrounded, first by a
light colored ring, the scleral ring, and externally to this by an irregular black stripe,
the chorioidal ring, which is especially well marked on the temporal side. The central artery and vein divide immediately after their entrance into the eye into an ascending and descending branch which appear somewhat lighter than their continu-
ations upon the retina, because they lie in the depth of the physiological excavation.
The branches, while still on the papilla, split into a number of smaller divisions and
fine off-shoots from them run from all directions toward the macula lutea, which itself is devoid of vessels, and is distinguished by its darker color. In its center a
bright punctate reflex, *f*, is visible.

FIGURE 55.   (After Fuchs.)

The Retina. The retina is not visible to the observer when the fundus is very light in color, but in
cases where the deposit of pigment is great, it appears as a cloud before the choroid. (The appear-

ance of the retinal vessels will be described in connection with the optic disk.)

Macula Lutea or Yellow Spot. The macula lutea is found about two diameters of the disk to the outer side of the papilla, and on a level with a horizontal line drawn through the disk slightly below its center. It is of about the same size as the optic disk and is darker in color than the rest of the fundus. The blood vessels of the macula lutea cannot be seen, but its circumference is bordered by minute branches of the retinal vessels, which are observable. At the center of the macula lutea, there is sometimes seen a small bright spot of a yellowish color. This is the *fovea centralis.* (Nettleship has demonstrated by injection, that numerous capillaries occupy all parts of the macula lutea with the exception of the fovea centralis, which is entirely free from these capillaries). There are two factors that contribute to the formation of the yellow spot: first, a thinning of Jacob's membrane at this point for the reason that the rods are wanting; second, the optic nerve fibres are lacking here.

The Optic Disk and Retinal Vessels. One of the best descriptions of the general appearances of the optic disk and retinal vessels in the healthy eye is given by C. Macnamara, F. R. C. S., in his "Diseases of the Eye," which was published in 1882. It is as follows: "The optic disk, or papilla, which is the termination of the optic nerve, or the spot at which it expands into the retina, will be found about one-tenth of an inch internal to the axis of the eye; it is the first point which attracts the observer's atten-

tion in making an examination with the ophthalmoscope. The shape of the healthy papilla is generally circular, but it frequently appears oval, because
the optic nerve and papilla are inserted sideways
into the eye, and we see it more or less obliquely,
and consequently, it is shortened in its horizontal
diameter. In other cases this oval form is due to a
real irregularity of the optic nerve, or to an irregularity in the dioptric media, notably in astigmatism.
The size of the optic disk is by no means the same in
all cases, and will appear to be augmented or les-
sened according to the power used to magnify it.

The color of the disk is not uniform, its outer
part being grayish and mottled. This appearance
is caused by the difference in the light reflected from
the nerve tubules, which is grayish, and that from
the white, glistening bands forming the lamina cribrosa. At the point of exit of the retinal vessels the
white appearance is very marked, and often presents
a little pit or hollow. The inner half of the disk is
of a decidedly redder tint than the outer half, because it is more thickly covered by vessels and nerve
fibres, and hence there is no reflection from the
fibres of the lamina cribrosa in this situation. It
is absolutely necessary to become acquainted with
the different appearances which may be presented
by the healthy optic disk, or these varying conditions may be mistaken for indications of disease;
the outer grayish-white tint, the central depressed
appearance and whitish hue, together with the inner
pinkish half of the disk, are conditions which vary

considerably, but are more or less distinctly recognizable in all healthy eyes.

At the point where the lamina cribrosa ceases, the optic nerve is contracted, and the opening in the choroid being narrow, in a certain measure compresses the nerve trunk; for this reason a sort of double border is often seen around the margin of the optic papilla. Under the choroidal margin is the line, more or less dark, that indicates the border of the opening in the choroid; under the sclerotic margin is a bright crescent or circle, formed by the curving round of the sclerotic fibres, and appearing between the choroidal margin and the fine grayish line that indicates the narrowest part of the nerve itself, and is therefore called the proper nerve boundary. The latter under normal circumstances, is not usually sharply defined. The choroidal rim is always strongly marked, especially at the outer border of the disk, where it sometimes has a well-defined deposit of pigment; this must not be mistaken for a diseased condition of the parts.

The point at which the central artery and vein of the retina enter the eye through the optic disk is subject to considerable variation. Generally the artery passes through the whitish and depressed center of the papilla, and, after emerging from the disc divides dichotomously, its branches ramifying in all directions toward the periphery of the retina; but the central artery may perforate the disc at any other point; not unfrequently one or two larger branches are noticed in the center of the papilla, while others pass through its circumference, perhaps close up to the scleral margin of the disc.

The apparent calibre of the vessels will vary with the magnifying power employed in observing them; practice alone will thus enable us to appreciate abnormal changes in the calibre of these vessels. One frequently reads accounts in which the retinal vessels are said to be over-full or empty, as the case may be; but in truth it is most difficult to determine this point.

The arteries, as well as their branches, are thinner, lighter in color, and straighter than the veins, which are darker in color and more sinuous in their course. The arteries seem to be transparent in their centers. This arises from the difference in the degree of illumination of the prominent centers of the arteries, as contrasted with their sides; from their conformation, it is evident that the sides of a vessel would receive and reflect relatively less light, and therefore appear in shade.

If in the normal eye the central vein be carefully examined, a pulsation may often be noticed in it, which will be rendered more evident on gentle pressure being made on the eyeball. If the compressing force be increased beyond a certain point, the pulsation at once stops, and the veins become almost invisible from the cessation of the flow of blood through them. In the healthy eye no arterial pulse can be seen, but if pressure be made on the eyeball it will become apparent. We noticed this in a marked manner in cases accompanied with considerable intra-ocular pressure, as for instance, in glaucoma.

The small depression on the surface of the disk is called the physiological excavation, in order to distinguish it from the pathological depression which is sometimes due to an abnormal intra-ocular pressure, as in glaucoma, or to atrophy of the optic nerve fibres."

CHAPTER XXIII.

THE perimeter is an instrument used to determine the dimensions of the field of vision.

The *field of vision* (visual field) is that portion of space containing all the points that are visible to the eye remaining fixed in one position.

Emerson's perimeter (a cut of which is given) is one of the best and is thus described:

FIGURE 70.—The Perimeter.

"The arc is a semicircle of 12.7 C. M. (5 inches) radius, revolving on a hollow spindle, and is divided on its convex surface into eighteen equal parts, numbered from the center to the extremities. On each arm of the arc is a perforated slide, so made that small pieces of paper can represent the objective point; in testing the color zones colored paper can be used. The arc is supported by a quadrant, mounted upon an adjustable upright set in a firm brass base. The scale on which the angle of revolution is measured is fixed to the quadrant, and a pointer attached to the revolving arc indicates the meridian tested. The chin rest is double, the right for the left eye and *vice versa*. The eye of the person tested should be 12.7 C. M. (5 inches) from the aperture and on a level with it."

The method of examination is as follows: Before making the examination it is necessary to bandage or cover the eye that is not being tested. The patient having placed his chin on the rest, he is directed to look at the fixation point which is situated at the midle of the semi-circle. We will suppose the semi-circle to be in a horizontal position. The perforated slide to the right is now moved slowly from the extremity towards the center of the semi-circle. At the instant the patient discerns the paper in the slide the point on the convex surface at which the slide stops is recorded on the perimeter chart.

In the same manner the slide to the left is moved toward the center and the point at which it is first discerned is recorded on the chart.

We now revolve the semi-circle a few degrees (say 30 degrees) and proceed as we did when the semi-circle was horizontal.

In like manner we revolve the semi-circle the same number of degrees each time and record the points at which the slide becomes visible. By connecting the points thus recorded on the chart, the field of vision is found.

Having obtained the field of vision of this eye, we proceed in like manner to find the visual field of the other eye.

Where a white paper is used in the slide, the field of vision is largest. Where the different colors are used, the field decreases in size in the following order: blue, yellow, orange, red and green, green having the most limited field.

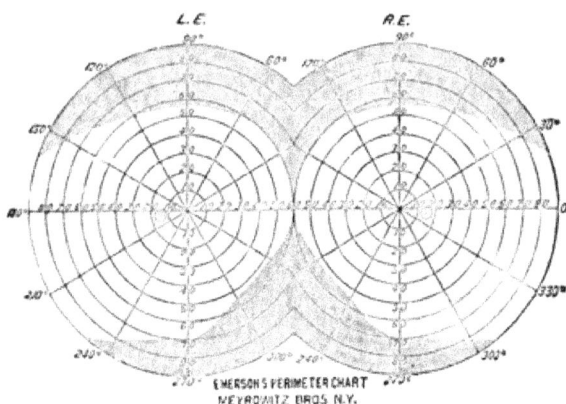

FIGURE 57.

On examining the accompanying charts, it is seen that the field of vision is not circular but of an irregular shape, extending upward about 50 degrees, outward somewhat over 90 degrees, inward 45 degrees, and downward 65 degrees. Owing to the projection formed by the nose inward and the supra-orbital margin above, the field of vision inward and upward is not as extensive as that of the temporal side.

The *blind spot* is the papilla which marks the entrance of the optic nerve and is so called because it cannot appreciate visual impressions.

Corresponding to the blind spot is a small island or *scotoma* in the field of vision situated 15 degrees to the outer (temporal) side of the fixation point.

The perimeter is of value in determining the size and position of a scotoma.

A scotoma is *positive* when it is apparent to the patient as a dark area in the field of vision.

In a *negative scotoma*, objects simply disappear when within the limits of the scotoma.

A *central scotoma* is one that occurs at the point of fixation and is due probably to a lesion of the optic nerve or macula.

Annular scotoma is one that surrounds the point of fixation and is circular.

Scotoma scintillans appears as a luminous cloud or mist before the patient's eyes. When it has an irregular outline like that of the wall of a fort, the condition is known as *teichopsia*.

The condition in which one-half of the field of vision is absent is called *hemianopsia*.

Temporal hemianopsia is that condition in which the temporal halves of both visual fields are absent.

Nasal hemianopsia is that condition in which the nasal halves of both visual fields are absent.

Homonoymous or *equilateral hemianopsia* is that condition in which the temporal half of one visual field, and the nasal half of the other are wanting.

PLACIDO'S DISC.

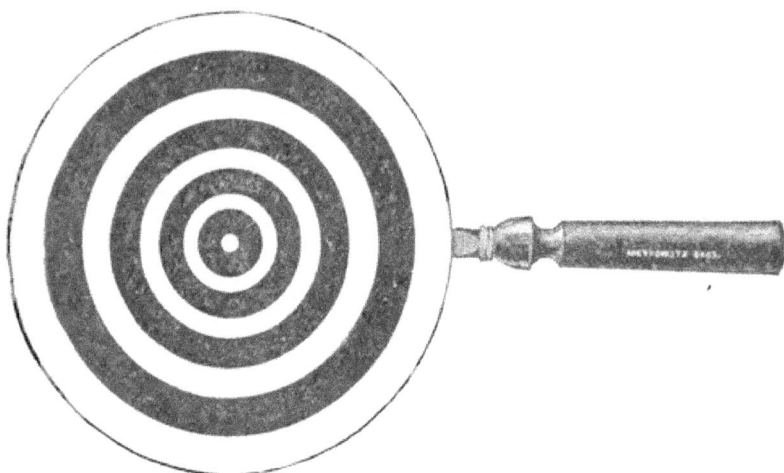

FIGURE 58.—Placido's Disc.

Placido's disc is an instrument used to determine in a short time whether astigmatism is present.

It consists of a disc about ten inches in diameter and has a central aperture which is used as a sight-hole for the observer. On the disc are concentric circles of a dark color, the background being white. The observer gazing through the sight-hole, holds the instrument about ten inches from the eye of the patient and sees the circles reflected on the cornea of the patient. If the circles are distorted, astigmatism is present. This is an exceedingly simple test and can be used by any physician.

OPHTHALMOMETRY.

The *ophthalmometer* is an instrument for determining the amount of astigmatism by examination of images reflected from the surface of the cornea.

The description of the instrument and the method of examination as given by Gertrude A. Walker, M. D., of Philadelphia, in her "Students' Aid in Ophthalmology," is one of the best, and is as follows: "A telescope is supported by an upright bar with a movable tripod base. Within the telescope at the focus of the eye-piece are two fine cross hairs; the telescope is also furnished with a

FIGURE 59.—Javal-Schiotz Ophthalmometer.

bi-refrigerant prism. To the large end of the telescope is attached a graduated arc, upon which are

two objects called targets, or mires, one of these (the
left) being fixed, while the other is movable.   Each
target is a parallelogram in shape, but one is cut
away in steps.   (See Fig. 60.)   At the outer side of
each target is a small pointer.   A much larger
pointer is attached to the telescope at about its
center.

The telescope passes through the center of
a large graduated disc.   Opposite the telescope
is a rest for the patient's head, and a small shade
which is used to cover the eye not under examination.   In testing, the ophthalmometer should be so

FIGURE 60.

placed that a strong light falls upon the disc.   The
patient's head having been placed in position, the
eye to be tested should look into the telescope.   The
observer now brings the patient's eye into the field
of the telescope and into focus by moving the tripod

FIGURE 61.

FIGURE 62.

base of the instrument backwards and forwards.
The targets and the disc are now reflected upon the

patient's cornea. Because of the prism in the telescope, there appear to the observer's eye four images of the targets. The outer ones are to be disregarded. The telescope should now be rotated so as to bring the long pointer to the zero mark on the disc. The left target is now brought into the center of the field, so that the cross hairs seem to divide it into four equal parts. Then the second target is made to slide along the arc until its edge seems just to touch, but not to overlap that of the first one. (See Fig. 61.)

The telescope should now be slowly rotated. If astigmatism be present, the target images will either separate or overlap. (See Figs. 60, 62.) The point of greatest separation or overlapping is found and the graduation upon the disc as indicated by the large pointer is noted. This expresses the direction of one of the principal meridians of the cornea. The extent of the separation or overlapping indicates the amount of astigmatism, each step of the target image being equivalent to one diopter. The extent of overlapping or separation is best found by noting the number of graduations that are passed over when the target is moved along just far enough to bring the edges of the two images together as at the beginning of the test. In case of overlapping, the small pointers indicate the meridian of greater curvature, while in case of separation they indicate the meridian of less curvature."

FIGURE 63.—Double Fixation Hook.

FIGURE 64.—Beers' Cataract Knife.

FIGURE 65.—Levis' Lens Scoop.

FIGURE 66.—Lens Scoop.

FIGURE 67.—Curved Cataract Needle.

FIGURE 68.—Knapp's Needle Cystotome.

FIGURE 69.—Dix's Spud.

FIGURE 70.—Beers' Cataract Needle.

FIGURE 71.—Stevens' Tenotomy Hook.

FIGURE 72.—Payne's Pterygium Knife.

FIGURE 73.—Gruening's Cautery Probe.

FIGURE 74.—Strabismus Hooks.

FIGURE 75.—Knapp's Cystotome.

FIGURE 76.—Keratomes.

FIGURE 77.—Paracentesis Needle.

FIGURE 78.—Stevens' Tenotomy Divulsor.

FIGURE 79.—Curved Scissors, Blunt Point.

FIGURE 80.—Stevens' Tenotomy Scissors.

FIGURE 81.—Curved Iris Scissors.

FIGURE 82.—Manoir's Scissors.

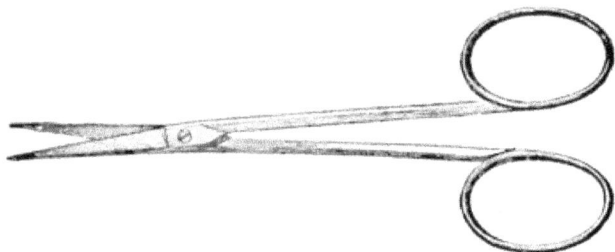

FIGURE 83.—Straight Iris Scissors.

FIGURE 84.—Noyes' Iris Scissors.

FIGURE 85.—Wecker's Iris Scissors.

FIGURE 86.—Liebreich's Iris Scissors.

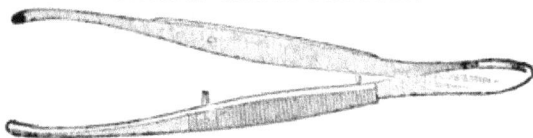

FIGURE 87.—Cilia Forceps.

FIGURE 88.—Fisher's Iris Forceps.

FIGURE 89.—Stevens' Tenotomy Forceps.

FIGURE 90.—Dressing Forceps.

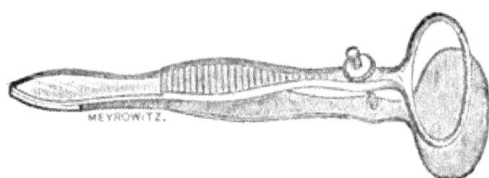

FIGURE 91.—Desmarre's Entropium Forceps.

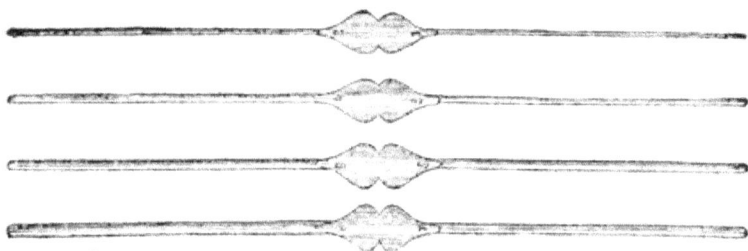

FIGURE 92.—Bowman's Lachrymal Probes.

FIGURE 93.—Stevens' Needle Holders.

FIGURE 94.—Desmarre's Lid Retractor.

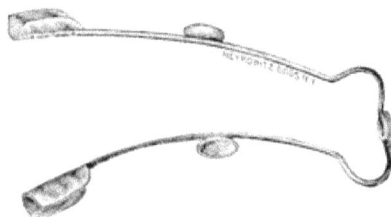

FIGURE 95.—Stevens' Tenotomy Eye-Speculum.

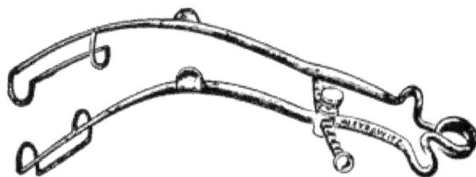

FIGURE 96.—Universal Eye-Speculum.

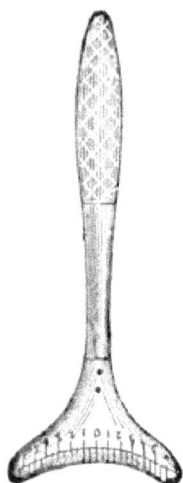

FIGURE 97.
Strabismometer.

FIGURE 98.
Andrews' Aseptic
Syringe for Anterior
Chamber.

FIGURE 99.
Aseptic Atropine
Dropper and Bottle.

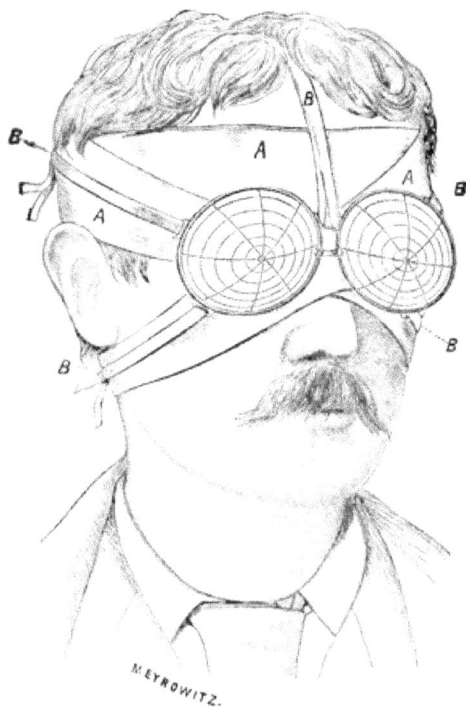

FIGURE 100.—McCoy's Aseptic Eye Shield.

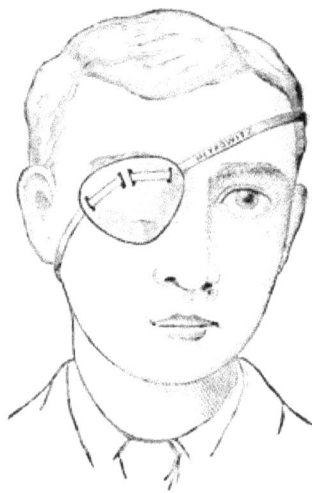

FIGURE 101.—Andrews' Eye Shield.

GLOSSARY.

ABDUCENS OCULI (*ab-du'-senz ok'-yu-li*). The rectus oculi externus muscle; also a nerve supplying the rectus oculi externus.

ABERRATION, (*ab-er-a'-shun*) OPTICAL. A scattering of the rays of light passing through a lens, so that they fail to unite at a focus.

ABLATIO RETINA (*ab-la'-she-o ret'ina*). A detachment of the retina.

ABLEPHARIA (*ah-blef-ar'-e-ah*). A total or partial absence of the eye-lids.

ABLEPSIA (*ah-blep'-se-ah*). Want of sight; blindness.

ABRASIO-CORNEA (*ab-ra'-ze-o kor'-ne-ah*). A shaving or scraping off of superficial opacities from the cornea.

ABRIN (*a'-brin*). The active principle of jequerity.

ABSUS (*ab'-sus*). A mixture composed of powdered cassia seeds and sugar, used in Egypt for the treatment of ophthalmia.

ACCOMMODATION (*ak-kom-mo-da'-shun*). The act by which the eye is adjusted for different distances.

Positive a., is the adjustment of the eye for near points.

Negative a., is the adjustment of the eye for distant points.

Region of a., is the linear distance between the far-point and the near-point.

Range of a., is the change in the refractive condition of the eye produced by the accommodation.

Spasm of a., is a continuous spasmodic cramp or contraction of the ciliary muscle, producing increased convexity of the crystalline lens, and making the eye appear to have a higher refractive power than it really possesses.

ACHLOROPSIA (*ah-klor-op'-se-ah*). Green blindness.

ACHNE (*ak'-ne*). A flake of mucus-like substance on the cornea.

ACHROMATIC LENSES (*ah-kro-mat'-ik lenses*). Those constructed of a combination of crown and flint glass, so arranged as to obviate chromatic aberration.

ACHROMOTOPSIA (*ah-kro-mat-op'-se-ah*). Total color blindness.

ACUS OPHTHALMIA (*a'-kus of-thal'-me-ah*). A couching or ophthalmic needle.

ACYANOBLEPSIA (*ah-si-an-o-blep'-se-ah*). The inability to distinguish blue.

ADACRYA (*ah-dak'-re-ah*). A deficiency of the lachrymal secretion

ADDUCENS OCULI (*ad-du'-senz ok'-yu-li*).   The rectus oculi internus.

ADENOPHTHALMIA (*ad-en-off-thal'-me-ah*).   An inflammation of the meibomian glands.

ADVANCEMENT (*ad-vans'-ment*).   Is an operation in which the tendon of a rectus muscle is brought forward to a new attachment.

AEGILOPS (*e'-ji-lops*).   A fistulous ulcer under the inner angle of the eye.

AFTER CATARACT (*aft'-er kat'-ar-akt*).   Portions of the capsule and lens remaining after extraction.

AGLIA (*ag'-le-ah*).   A whitish speck on the cornea.

AKYANOPSIA (*ah-ki-an-op'-se-ah*).   Violet blindness.

ALBUGO (*al-bu'-go*).   A white opacity of the cornea.   The same as leucoma.

ALLOCHROMASIA (*al-o-kro-ma-ze'-ah*).   A difference or change in color.

ALTERNATING SQUINT (*awl'-ter-nat-ing squint*).   The condition obtaining when the two eyes squint alternately.

AMAUROSIS (*am-aw-ro'-sis*).   A loss of sight without perceptible ocular lesions.

AMBLYOPIA (*am-ble-o'-pe-ah*).   A dimness of sight; particularly weak vision unaccompanied by organic changes in the eye, and not benefitted by glasses.

AMETROMETER (*ah-met-rom'-et-er*).   An instrument for measuring ametropia.

AMETROPIA (*ah-met-ro'-pe-ah*).   A condition of the eye in which the refracting powers of the media are not adjusted to the position of the retina.

ANERYTHROPSIA (*an-er-ith-rop'-se-ah*).   Red blindness.

ANGLE-ALPHA (*ang'-gel al'-fa.*)   The angle formed by the line of vision with the major axis of the corneal ellipse.

> *Gamma* a., the angle formed by the line of fixation with the axis of the eye.
>
> *Convergence,* a. of, measured by the angle through which an eye turns when it abandons parallelism to fix a near object.
>
> *Incidence,* a. of.   The angle made by an incident ray with the perpendicular.
>
> *Refraction,* a. of.   The angle formed by the refracted ray with the perpendicular.
>
> *Vision,* a. of, is the angle between the two lines drawn from either extremity of an object to the eye.

ANIRIDIA (*an-ir-id'-e-ah*).   Absence of the iris.

ANISOCORIA (*an-is-o-ko'-re-ah*).   Inequality of the pupils.

ANISOMETROPIA (*an-is-o-me-tro-pe-ah*).   The state in which the refraction of the two eyes is unequal.

ANKYLOBLEPHARON (*ang-kil-o-blef'-ar-on*) A total or partial adhesion of the borders of the lids.

ANNULAR MUSCLE OF MULLER (*an'-u-lar mus'-l of Mueller*). A portion of the ciliary muscle.

ANOPHTHALMOS (*an-off-thal'-mus*). Absence of the eye.

ANOPSIA (*an-op'-se-ah*). Defect of sight; especially poor sight due to defectiveness of the eye.

ANORTHOPIA (*an-or-tho'-pe-ah*). A natural defect of sight in which one is unable to detect a want of symmetry.

ANTERIOR CHAMBER (*an-te'-re-or cham'-ber*). The space between the posterior surface of the cornea and the anterior surface of the lens.

ANTIMETROPIA (*an-te-met-ro'-pe-ah*). That condition in which one eye is hyperopic and the other is myopic.

APHAKIA (*ah-fa'-ke-ah*). The condition of an eye without the crystalline lens.

APLANATIC (*ah-plan-at'-ic*). Unaffected by spherical aberration.

AQUEOUS HUMOR (*a'-kwe-us u'-mor*). A colorless fluid in the anterior and posterior chambers of the eye.

AQUULA (*ak-zu'-lah*). An aqueous or fatty tumor under the skin of the eye-lids.

ARCUS SENILIS (*ar'-kus se-ni-lis*). A narrow gray line of degeneration which runs around the cornea; due to deposition of colloid material.

ARGYRIA (*ar-jir'-e-ah*). A deposition of silver oxide in the tissues of the conjunctiva.

ARTERY HYALOIDEA (*ar'-ter-e hi-al-oid'-e-ah*). The artery which nourishes the vitreous and lens in the fetus. It runs antero-posteriorly through the vitreous humor. It disappears usually after birth but in some instances persists.

ARTIFICIAL EYE (*ar-te-fish'-al i*). An eye made of glass or celluloid.

ARTIFICIAL PUPIL (*ar-te-fish'-al pu'-pil*). The result of an operation for overcoming the effect of adhesions or permanent contraction of the iris.

ASTHENOPIA (*as-then-o'-pe-ah*). Rapid tiring of the eyes upon exertion, manifested by a sense of pain in the eyes, headache, amblyopia, etc.

    Accommodative a. Is due to fatigue of the ciliary muscle owing to excessive strain required by the presence of a refractive error, as hyperopia or astigmatism.

    Muscular a. Is due to insufficiency or weakness of the muscles of the eye.

    Nervous a. Is due to central causes, such as hysteria.

ASTIGMATISM (*as-tig'-mat-ism*). A condition in which the refractive power of the eye varies in the different meridians, so that the rays which enter it along one meridian are brought to a focus sooner than those which enter it along another. This condition makes lines running in all directions appear different, although they are alike.

Simple a., is that condition of the eye in which it is emmetropic in one diameter and hyperopic or myopic in the other meridians.

Compound a., is that condition of the eye in which there is hyperopia or myopia in all meridians, but more so in some than in others.

Mixed a., is that condition of the eye in which there is hyperopia in some meridians and myopia in others.

Regular a. is that form of astigmatism in which is found one meridian of greatest and one of least refraction, which two meridians usually lie at right angles to each other, which meridians are known as principal meridians.

Irregular a. is that form in which the unequal curvatures of the cornea bear no constant relation to each other; therefore there are no principal meridians.

Corneal a. is caused by irregularity of the curvature of the cornea.

Lenticular a. is caused by an irregularity of the curvature of the lens.

ASTIGMOMETER (*as-tig-mom'-et-er*). An instrument for locating and measuring astigmatism.

ATROPINE or ATROPIN (*at'-ro-pin*). An alkaloid of belladonna used as a mydriatic.

AUTOPHTHALMOSCOPY (*aw-toff-thal'-mo-skop-e*). The use of the ophthalmoscope on one's self.

BASEDOW'S DISEASE (*Ba'sedow's dis-ez'*). Exophthalmic goitre. It is so called from Basedow, who described it in 1840.

BINOCULAR (*bin-ok'-u-lar*). With, or by means of both eyes, as binocular vision.

"BLACK EYE" (*black-i*). Ecchymosis of the eye-lids.

BLEAR EYE. A chronic inflammation of the eye lids. Marginal blepharitis.

BLENNOPHTHALMIA (*blen-off-thal'-me-ah*). An inflammation of the mucous membrane of the eye accompanied by a purulent or muco-purulent discharge.

BLEPHARADENITIS (*blef'-ar-ad-en-i-tis*). An inflammation of the meibomian glands.

BLEPHARITIS (*blef-ar-i'-tis*). An inflammation of the eye-lids.

BLEPHARITIS MARGINALIS (*blef-ar-i-tis marj-in-al'-is*). (Blepharadenitis; blepharitis ciliaris) a chronic diffuse inflammation of the sebaceous glands along the margins of the lids.

BLEPHARO–ADENITIS (*blef'-ar-o–ad-en-i-tis*). Same as blepharadenitis; blepharitis marginalis.

BLEPHARO–ADENOMA (*blef-ar-o–ad-en-o-mah*). An adenoma of the margin of the lids.

BLEPHAROCHROMIDROSIS (*blef'-ar-o-kro-mid-ro'-sis*). Pigmentation of the lids occuring in spots upon the skin.

BLEPHARODOEMA (*blef-ar-o-e-de'-mah*). A watery swelling of the eye-lids.

BLEPHARONCUS (*blef-ar-ong'-kus*). A tumor on the eye-lid.

BLEPHAROPHIMOSIS (*blef-ar-o-fi-mo'-sis*). A congenital contraction of the palpebral fissure.

BLEPHAROPHTHALMIA (*blef-ar-off-thal'-me-ah*). An inflammation of the eye and the eye-lids.

BLEPHAROPLASTY (*blef'-ar-o-plas-te*). An operation for repairing any lesion of the lids by taking a flap from the contiguous parts.

BLEPHAROPLEGIA (*blef-ar-o-ple'-je-ah*). The falling down of the upper lid from paralysis.

BLEPHAROPTOSIS (*blef-ar-op-to'-sis*). Ptosis; a falling of the lid.

BLEPHARORRHAPHY (*blef-ar-or'-a-fe*). The operation of stitching together the upper and lower lids.

BLEPHAROSPASM (*blef'-ar-o-spazm*). A spasmodic contraction of the lids.

BLEPHAROSTAT (*blef'-ar-o-stat*). An eye speculum.

BLEPHAROTOMY (*blef-ar-ot'-o-me*). Cutting of the orbicularis.

BLIND SPOT (*blind spot*). The papilla. The entrance of the optic nerve.

"BLUE STONE" (*blue stone*). The sulphate of copper. Formerly much used in the treatment of granular ophthalmia.

BRACHYMETROPIA (*brak-e-me-tro'-pe-ah*). Same as myopia.

BUPHTHALMIA (*buf-thal'-me-ah*). A marked enlargement of the eye in all of its parts. It is a disease of childhood and is probably glaucomatous.

CALIGO (*kal-i'-go*). Dimness of sight, or blindness, sometimes coming on without apparent cause.

CALIGO CORNEA (*kal-i'-go kor'-ne-ah*). Dimness of sight from opacity of the cornea.

CALIGO HUMORUM (*kal-i'-go u'-mor-um*). An obscurity of vision arising from a defect in the humors of the eye.

CALIGO LENTIS (*kal-i'-go len'-tis*). An opacity of the crystalline lens. The true cataract.

CAMERA (*kam'-er-ah*). The anterior and posterior chambers of the eye.

CANALICULIS (*kan-al-ik'-u-lus*). A canal for the passage of tears from the lachrymalis to the lachrymal sac.

CANAL OF CLOQUET (*kan-al' of Klo-ka*). A channel for the hyaloid artery.

CANAL OF SCHLEMM (*kanal' of Schlemm*). A venous space at the junction of the sclerotic and cornea.

CANITIES (*kan-ish'-e-ez*). Decoloration of the lashes.

CANTHOPLASTY (*kan'-tho-plas-te*). The operation of transplanting a portion of the the occular conjunctiva to the external canthus. Extension of the canthus by any operation.

CANTHOTOMY (*kan-thot'-o-me*). The operation for enlarging the palpebral fissure.

CANTHUS (*kan'-thus*). The angle formed by the junction of the eye-lids.

CAPSULE OF TENON (*kap'-sul of Te'-non*). A delicate fibrous sheath enveloping the eye-ball and forming the socket in which the globe revolves.

CAPSULITIS (*kap-su-le'-tis*). Inflammation of the capsule (Tenon's) covering the eye.

CAPSULOTOME (*kap'-su-lo-tom*). An instrument for incising the capsule of the crystalline lens.

CAPSULOTOMY (*kap-su-lot'-o-my*). The operation of incising the capsule of the crystalline lens as in operations for cataract.

CARUNCULA (*kar-un'-ku-lah*) lachrymalis. The small red body situated in the inner angle of the eye.

CATACLEISIS (*kat-ak-li'-sis*). A morbid or spasmodic closing of the eye-lids

CATAPHORIA (*kat-af-o'-re-ah*). A tending of the visual line downward.

CATARACT (*kat-ar-akt*). Obstructed vision caused by opacity of the crystalline lens or its capsule.

Capsular c., an opacity upon the capsule of the lens.

Lenticular c., an opacity of the lens proper.

Senile c., opacity of the lens due to age.

Traumatic c., cataract due to injury.

CATOPTRICS (*kat-op'-triks*). That branch of optics which treats of the reflection of light.

CATOPTRIC TEST (*kat-op-trik test*). A test which depends upon the three images seen upon the healthy eye from a light held before it.

CENTERING OF LENSES (*sen'-ter-ing of lenz'-es*). The adjusting of lenses so that the optical centre of the glass is exactly in front of the pupil.

CENTRAD (*sen'-trad*). A prism which will produce a deviation equal to one-hundredth of a radian.

CENTRA DIAPHANES (*sen'-trah di'-af-anes*). Cataract caused by obscurity of the central portion of the lens.

CENTRE OF CURVATURE (*sen'-ter of ker'-vat-ur*). That point through which rays pass without being reflected.

CENTRE OF ROTATION (*sen'-ter of ro-ta'-shun*). The point about which the eye revolves.

CERATOME. A knife for dividing the cornea.

CERATOMY. Cutting of the cornea.

CHALAZION (*kal-a-ze'-on*). A tumor of the lids formed by distention of one of the meibomian glands.

CHEMOSIS (*ke-mo'-sis*). Swelling and oedema of the ocular conjunctiva.

CHOKED DISK. A swelled or oedematous condition of the optic disk occurring as a result of increased intra-cranial pressure.

CHORIO-CAPILLARIS (*ko-re-o-kap-il-a'-ris*). The inner of the three vascular layers of the choroid, consisting of a network of capillary vessels.

CHOROID (*ko'-roid*). The posterior segment of the uvea, or middle tunic of the eye.

CHOROIDITIS *ko-roid-i'-tis*). Inflammation of the choroid.

CHOROIDO-IRITIS (*ko-roid'-o-i-ri'-tis*). An inflammation of the choroid and iris.

CHOROIDO-RETINITIS (*ko-roid'-o-ret-in-i'-tis*). An inflammation of the choroid and retina.

CHROMATISM (*kro'-mat-izm*). The prismatic aberation of the rays of light.

CHROMATOGENOUS (*kro-mat-of'-en-us*). Generating color or pigment.

CHROMATOLOGY (*kro-mat-ol'-o-ji*). The science of colors.

CHROMATOPSY (*kro'-mat-op-se*). Colored vision.

CHROMATROPE (*kro'-mah-trop*). An instrument for exhibiting a variety of colors, producing, by a rapid revolving motion, beautiful pictures.

CILIA (*sil'-e-ah*). The eye-lashes.

CILIARY BODY (*sil'-e-a-re body*). The mid portion of the uvea or pigmentary tunic of the eye; composed of the ciliary muscle and the ciliary processes.

CILIARY MUSCLE. The circular muscle of accommodation of the eye.

CILIARY NERVES. *Long*, branches of the nasal which supply the ciliary muscle.
Short, nerves from the ciliary ganglion which supply the ciliary body.

CILIARY NEURALGIA. Irritation of the ciliary nerves characterized by pain in and around the eye, over the brow, and down the side of the face.

CILIARY REGION. Of or pertaining to the ciliary body; that portion of the globe which is concerned in accommodation.

CILLOSIS (*sil-o'-sis*). Spasmodic trembling or agitation of the eye-lids.

CIRCLE OF DIFFUSION. The image of the retina formed by a near object when the eye is adjusted for distance.

CIRCULUS ARTERIOSUS IRIDIS. An artery which encircles the iris.

CIRCUMAGENTES (*sir-kum-aj'-en-tez*). The oblique muscles of the eye.

CIRCUM-CORNEAL ZONE. A pink zone of vessels around the cornea.

CIRCUM-LENTAL SPACE. The space between the equator of the lens and the ciliary processes.

CIRSOPHTHALMIA (*sir-soff-thal'-me-ah*). A varicose condition of the eye.

CLAIRVOYANCE (*klar-voy-anz*). Literally, "clear-sightedness," or "clear vision."

COCAINE (*ko-kah'-in* or *ko'-kan*). An alkaloid obtained from the leaves of the Erythroxylon coca.

COLLYRIUM (*kal-ir'-e-um*). An eye-wash; a medicated application, usually a lotion for the eyes.

COLOBOMA (*kol-o-bo'-mah*). A gap or fissure, usually congenital, in any part of the eye or the eye-lid.

COLOR. The impression which the light reflected from the surface of bodies makes on the organs of vision.

COLOR-BLINDNESS. The inability to recognize colors correctly.

COLOR SENSE. The power which the retina has of perceiving color.

COMMOTIO RETINA (*kom-o'-she-o retina*). Oedema of the retina.

CONCAVO-CONVEX. Having one face concave, the other convex.

CONCOMITANT SQUINT. A form of squint in which one eye, although deviated, moves in conjunction with the other, so that the amount of deflection remains the same in all parts of the field of vision.

CONICAL CORNEA (keratoconus). A condition of the cornea in which it bulges forward in the form of a cone.

CONJUGATE FOCUS. Any other focus besides the principal focus.

CONJUGATE PARALYSIS. Loss of power of motion of the two eyes in some one direction.

CONJUNCTIVA (*kon-junk-ti'-vah*). The mucous membrane lining the eye-lids and eye-ball.

CONJUNCTIVITIS (*kon-junk-tiv-i'-tis*). An inflammation of the conjunctiva.

CONUS (*konus*). The wedged shaped posterior staphyloma found in the fundus of the eye in myopia. Also called the myopic crescent.

CONVERGENCE. The act of converging or of being directed toward a common point. In vision, the direction of the lines of fixation so that both fall upon the point fixed.

CONVERGENT STRABISMUS. The turning in of the eye-ball.

COPIOPIA (*kop-e-o'-pe-ah*). Fatigue or weariness of vision.

COQUILLES (*ko'-kil*) (Fr.) Shelled shaped glasses.

CORE. The pupil.

CORECLEISIS or COROCLISIS (*ko-ro-kli'-sis*). Obliteration of the pupil.

CORECTOMY (*kor-ek'-to-me*). The operation for artificial pupil by cutting away part of the iris.

CORECTOPIA (*kor-ek-to'-pe-ah*). An eccentric position of the pupil.

COREDIALYSIS (*kor-e-di-al'-is-is*). A separation of the iris from its attachment.

CORELYSIS (*kor-el'-is-is*). The operation of breaking adhesions between the iris and lens.

COREMORPHOSIS (*kor-e-mor'-fo-sis*). The operation of making an artificial pupil.

CORENCLEISIS (*kor-en-kli'-sis*). An operation for artificial pupil.

COREOMETER (*kor-c-om'-et-er*). An apparatus for measuring the width of the pupil.

COREONCION (*kor-e-on'-se-on*). A kind of hook for the operation of artificial pupil.

COREPLASTY (*kor-e'-plas-te*). Operations in general for artificial pupil.

CORNEA (*kor'-ne-ah*). The transparent convexo-concave substance forming the anterior part of the eye-ball.

CORNEAL ASTIGMATISM. That condition in which the radii of curvature of the cornea are not equal.

CORNEAL LOUPE. A strong, mounted lens, for examination of the cornea by oblique illumination.

CORNEAL OPACITY. The scar remaining after a lesion of the cornea.

CORNEAL REFLEX. Illumination of the cornea for the purpose of determining the presence of ametropia.

CORNEAL ULCER. An ulcer on the surface or in the substance of the cornea.

CORNEITIS (*kor-ne-i'-tis*). Inflammation of the cornea.

CORPUS VITRIUM. The vitreous body.

CORRUGATOR SUPERCILII. The muscle that wrinkles the brow.

CORTEX OF LENS. The outer part or shell of lens.

CORTICAL CATARACT. Opacity of the cortical layers of the lens.

COUCHING (*kowch'-ing*). An old operation for cataract consisting in depressing the lens into the bottom of the vitreous.

CROSSED DIPLOPIA. A tending of the axis of the eye outward, thus projecting the line of the vision across that of the other eye.

CROUPOUS OPHTHALMIA. Membranous exudation and soft, painless swelling of the conjunctiva.

CRUSTA LACTEA. A moist eczema of the lids of children.

CRYPT-OPHTHALMUS. Absence of both eye-lid and conjunctival sac.

CRYSTALLINE LENS. A transparent, double convex lens, situated in the fore part of the vitreous humor of the eye.

CUL-DE-SAC (*kul'-de-sak'*). The fold of transition between the ocular and palpebral conjunctiva.

CUP. A depression of the optic disk in glaucoma.

CYCLITIS (*sik-li'-tis*). Inflammation of the ciliary body.

CYCLOPIA (*si-klo'-pe-ah*). Fusion of the two orbits in the middle of the face.

CYCLOPLEGIA (*si-klo-ple'-je-ah*). Paralysis of the ciliary muscle.

CYCLOTOMY (*sik-lot'-o-me*). Division of the ciliary muscle; an operation for glaucoma.

CYLINDERICAL LENS. A lens made from the segment of a cylinder.

CYSTICERCUS (*sis-te-ser'-cus*). A tailed worm, of the genus entozoa, sometimes found in the vitreous.

DACRYADENALGIA (*dak-re-ad-en-al'-je-ah*). Pain in the lachrymal gland.

DACRYADENITIS (*dak-re-ad-en-i'-tis*). Inflammation of the lachrymal gland.

DACRYOCYST (*dak'-re-o-sist*). The lachrymal sac.

DACRYOCYSTALGIA (*dak-re-o-sis-tal'-je-ah*). Pain in the lachrymal sac.

DACRYOCYSTITIS (*dak-re-o-sis-ti'-tis*). Inflammation of the lachrymal sac.

DACRYO-CYSTO-BLENNORRHŒA. A discharge of mucus from the lachrymal sac.

DACRYO-CYSTO-PYORRHŒA. A discharge of pus from the lachrymal sac.

DACRYOLITE (*dak'-re-o-lit*). A calcarious concretion in the lachrymal passage.

DACRYOMA (*dak'-re-o-mah*). An obstruction in the puncta lachrymalis, causing an overflow of tears.

DACRYOPEUS (*dak-re-o'-pe-us*). Causing tears.

DACRYOPS (*dak'-re-ops*). A cyst filled with clear liquid, due to the distention of one of the ducts of the lachrymal gland.

DACRYORRHŒA (*dak-re-or-e'-ah*). A morbid flow of tears.

DACRYOSOLEN (*dak-re-o-so'-len*). The lychrymal canal or duct.

DACRYOSOLENITIS (*dak-re-o-so-len-i-tis*). Inflammation of the lachrymal duct.

DALTONISM (*dal'-ton-izm*). Color blindness.

DANGER ZONE. The ciliary zone; that part of the sclerotic over the ciliary body.

DARK ROOM. A room darkened for the purpose of examination of eyes by artificial illumination.

DASYMA. Roughness on the internal surface of the eye-lid.

DATURINE (*da tew'reen*). An alkaloid derived from the datura stramonium, which causes dilatation of the pupil.

DECENTERED LENSES. Lenses in which the optical centre is not before the pupil.

DENDRIFORM (*dend'ri-form*) ULCERS. Dendriform keratitis; a kind of ulcer of the cornea, which has branch-like ramifications.

DEPLUMATION (*dep'lu-ma-shun*). A term applied to a disease of the eye-lids, in which the eye-lashes fall off.

DEPRESSIO CATARACTÆ (*de-presh'c-oh kat-a-rak'ti*). Artificial luxation of the lens. The oldest operation for cataract.

DEPRIMENS OCULI (*de-pri'mens ok'u-li*). A name given to the rectus inferior, from the action of this muscle in drawing down the eye-ball.

DESCEMETITIS (*des'e-me-ti'tis*). Inflammation of the membrane of Descemet.

DESCEMETOCELE (*des'e-me'to-seel*). A watery tumor of Descemet's membrane.

DESCEMET'S (*des-e-mayz*) MEMBRANE. The structureless, transparent membrane lining the posterior surface of the cornea.

DIABETIC (*di-a-bet'ik*) CATARACT. Cataract due to diabetes mellitus.

DILATOR PUPILLÆ (*di-lay'tur pew'pil-i*). The radiating fibres of the iris.

DIOPTOMETRY (*di-op-tom'e-try*). The determination of the refraction and accommodation of the eye.

DIOPTRE (*di-op'tur*). The unit used in measuring glasses and the refractive states of the eye.

DIOPTRIC (*di-op'trik*) APPARATUS OF THE EYE. The cornea, aqueous humor, crystalline lens, and vitreous humor.

DIPHTHERITIC OPHTHALMIA (*dif-thur-it'ik of-thal'me-ah*). A conjunctivitis produced by diphtheria. It is characterized by a hard, painful swelling of the lids, a scanty sero-purulent or saneous discharge, etc.

DIPHTHERITIC PARALYSIS. A paralysis of accommodation sometimes following diphtheria.

DIPLOPIA (*di-ploh'pee-ah*). Double vision. The condition in which an object seen appears double.

DIRECT IMAGE. The image as seen with the ophthalmoscope by the direct method.

DISC (*disk*) OPTIC. The entrance of the optic nerve.

DISCISSION (*dis-sizh'un*). The needle operation for cataract.

DISPERSING LENS. A concave lens.

DISSEMINATED CHOROIDITIS. An inflammation of the choroid in which the exudate is scattered generally over the fundus.

DISTICHIASIS (*dis-tik'e-a'-sis*). A condition in which there are two rows of cilia, one or both of which are turned inward toward the eye-ball.

DIVERGENCE. An outward deviation of the visual lines. The amount, normal or abnormal, which the eyes can roll outward from the position of fixation.

DIVERGENT SQUINT. Strabismus in which the eye turns out.

DOLICHOCEPHALY (*dol'i-ko-sef-aly*). A prominence of the bones of the face and orbit.

DUBOISIA (*dew-boy'she-ah*). The active principle of the duboisia myoporoides.

DUCTUS LACHRYMALIS (*dukt'us lak'ri-mal'us*). The duct leading from the lachrymal sac to the nasal fossa.

DYNAMIC (*di-nam'ik*) REFRACTION. The act of accommodation.

DYSCHROMATOPSIA (*dis-kroh'ma-top'see-ah*). Difficulty in distinguishing colors.

DYSCORIA (*dis-ko're-ah*). Non-circular form of the pupil.

DYSLEXIA. Hysterical asthenopia.

DYSOPSIA (*dis-op'se-ah*). Painful vision.

DYSOPSIA LATERALIS. An affection in which an object can only be seen when seen obliquely.

ECARTEURS (*a-kar'-tuze*). An instrument for separating the lids.

ECHINOPHTHALMIA (*e-kin-of-thal'-me-ah*). A form of ophthalmia in which the lashes project like the quills on a hedgehog.

ECTOPIA (*ek-toh'pe-ah*) OF LENS. A term applied to any case where the lens is out of its place. A dislocation of the lens, either congenital or traumatic.

ECTROPION (*ek-troh'-pe-an*). A turning out or eversion of the eye-lids. Same as Ectropium.

EGYPTIAN OPHTHALMIA. An acute form of ophthalmia, attended with a purulent secretion. Also called Military Ophthalmia.

EMMETROPE. Is one, whose eyes, when their accommodation is relaxed, are accurately adjusted for parallel rays.

EMMETROPIA (*em'-e-troh'-pe-ah*). The condition of the eye in which parallel rays will focus upon the retina without accommodation.

ENCANTHIS. A small red excrescence on the caruncula lachrymalis and the semilunar fold of the conjunctiva.

ENCANTHIS BENIGNA. Benign new growths of the caruncle.

ENCANTHIS MALIGNA. Malignant new growths of the caruncle.

ENOPHTHALMUS (*en'-of-thal'mus*). A condition in which the eye is unusually deep in its socket. A retraction of the eyeball.

ENTROPION (*en-troh'-pi-un*). That condition in which the eyelash and eyelid are turned in towards the eye. Same as Entropium.

ENUCLEATION (*e-new'-kle-a'shun*). The removal of the eye.

EPHIDROSIS (*ef'-e-droh'sis*). An excessive secretion of the sudoriferous glands of the upper lid.

EPICANTHUS (*ep'-e-kan'thus*). A condition in which a fold of skin projects over and hides the inner canthus of the eye.

EPILATION (*ep'e-lay-shun*). The removing of the cilia by plucking ing them out by the roots.

EPIPHORA (*e-pif'-ur-ah*) A condition in which the tears run over the lids instead of through the natural passage, due usually to stricture of the lachrymal passage.

EPISCLERITIS (*skle-ry'-tis, skle-re'-tis*). An inflammation of the ocular sub-conjunctival connective tissue.

ERROR OF REFRACTION Any deviation in the normal refraction of the eye. Ametropic.

ERYTHROPSIA (*er'-e-throp'-see-ah*). Red vision.

ESERINE (*es'-ur-een*). An alkaloid obtained from the calabar-bean, used as a myotic.

ESOPHORIA (*es'-oh-foh'-re-ah*). A tendency of the visual lines inward; an excessive convergence of the internal recti, owing to insufficiency of the externi.

ESOTROPIA (*es'-oh-troh'-pe-ah*). A manifest turning inward of the eyes; an inward or convergent squint.

EVISCERATION (*e-vis'-ur-a-shun*). The operation of removing the contents of the globe, the sclerotic being retained.

EXCLUSION OF PUPIL. The condition caused by the iris being bound down to the capsule of the lens.

EXENTERATION (*eks-en'-tur-a-shun*) The same as Evisceration.

EXOPHORIA (*eks'-o-foh'-re-ah*). A tendency of the eyes to deviate outward, so that they diverge from the point of fixation. Also termed insufficiency of the interni.

EXOPHTHALMIC (*of-thal'-mik*). Pertaining to exophthalmus; as exophthalmic goitre.

EXOPHTHALMIC GOITRE (*goy'-tur*). A disease, one of the symtoms of which is protusion of the eye-balls.

EXOPHTHALMOMETER (*mom'-e-ter*). An instrument for measuring the degree of exophthalmus.

EXOPHTHALMOS (*of-thal'-mos*). A condition in which the eye protudes abnormally from the socket, no matter what the cause.

EXTERNAL RECTUS. The muscle that turns the eye-ball outward.

EXTRINSIC MUSCLES. The muscles outside the globe.

EYE. The organ of vision.

EYEBROW. The fold of skin, lined with hairs, situated at the upper margin of the orbit.

EYELASH. The delicate hairs projecting from the edges of the eyelids.

EYELID. The projecting folds from above and below which cover the eye.

EYE-SPECULUM. An instrument for keeping the lids apart during an operation.

EYE WATER. A medicated eye wash.

FALSE IMAGE. The image formed by the squinting eye.

FAR-POINT. The farthest point at which, with full relaxation of accommodation, objects can be seen distinctly.

FAR SIGHTEDNESS. A defect of vision, by which objects cannot be seen perfectly, owing to the visual axis being too short, or to the hardening of the crystalline lens consequent to age.

FIBRES OF MULLER (*muel'ler*). Fibres of connective tissue which run perpendicularly through the retina.

FIELD OF VISION (VISUAL FIELD). The portion of space containing all the points that are visible to the eye, remaining fixed in one position.

FIFTH NERVE (THE TRIGEMINUS). The nerve supplying sensation to the face and mobility to the muscles of mastication. The ophthalmic nerve is derived from this.

FILARIA (*fy-lay're-ah*). A thread like parasitic worm which infests the cornea of the eye, particularly that of the horse.

FIXATION FORCEPS. Small forceps for steadying the eye in operations.

FLOCCI VOLITANTES (*flok'si vol'-i-tan'tez*). The imaginary objects floating before the eyes in cases of depraved sight. See Muscæ Volitantes.

FLUORESCEIN (*flew'o-res'e-in*). A coal-tar derivative which colors green those portions of the cornea which have been deprived of epithelium.

FOCAL ILLUMINATION. Examination of the eye with light focused upon it by a strong lens.

FOCAL INTERVAL. The interval between the focus of the meridian of greatest curvature and the focus of the meridian of least curvature of an irregular refracting surface.

FOCUS. The point at which rays meet after passing through a convex lens.

FOLD, SEMILUNAR. A crescentric fold of the conjunctiva at the inner canthus.

FOLLICULAR (*fo-lik'yu-lur*) OPHTHALMIA. A form of conjunctivitis characterized by small pinkish prominences on the conjunctiva.

FONTANA'S (*fon-tah'nah*) SPACES. The open spaces of the meshwork of the ligamentum pectinatum.

FORAMEN (*fo-ray'men*) SUPRA-ORBITAL. The supra-orbital hole or notch for the passage of the superciliary artery, vein and nerve.

FOREIGN BODY. Any foreign substance within the lids or within the globe.

FORNIX. The conjunctival cul-de-sac.

FOSSA LACHRYMALIS. A depression in the frontal bone for the reception of the lachrymal gland.

FOSSA PATELLARIS (*pa'tel-la'ris*). The depression in the front part of the vitreous for the lens.

FOVEA (*foh've-ah*) CENTRALIS. A small depression in the macula lutea.

FRANKLIN GLASSES. Same as bifocals.

FULMINATING GLAUCOMA. The most malignant type of glaucoma.

FUNDUS OCULI. The back part of the eyeball or the portion covered by the retina.

FUSION POWER. The power of the ocular muscles to make the two images fuse.

GERONTOXON (*jer'-on-toks'-on*). A whitish ring occuring in old people in the cornea, near, and concentric with its margin. Called arcus senilis. It is due to a deposition of colloid material.

GLASSES. Lenses made of glass to aid in vision.

GLAUCOMA (*glaw-ko'-mah*). An increase of intra-ocular pressure, producing hardness of the eye-ball, and rapidly failing vision, due to injury of the retina inflicted by pressure upon it.

GLAUCOMATOUS (*glaw-ko'-ma-tus*) RING. The yellowish halo seen around the excavated nerve head in glaucoma.

GLIOMA (*gly-oh'-mah*) RETINAE. A round cell sarcoma of the retina, occurring in young people.

GLOBE. The eyeball.

GOGGLES. Colored glasses with wire or silk sides to protect the eyes.

GONORRHŒAL (*gon'-ur-e'-al*) OPHTHALMIA. An acute purulent ophthalmia caused by an infection from a gonorrhœal discharge.

GONORRHŒAL IRITIS. A plastic iritis, said by some, to be due to gonorrhea.

GRANULAR LIDS. A granulated condition of the eyelids; known also, as trachoma.

GRAVES DISEASE. A disease which is characterized by an abnormal protuberance of the eyeballs.

GREEN. The fourth color in the spectrum.

GYMNASTIC PRISMS. Prisms placed before the eyes for the purpose of strengthening the muscles of the eyes by exercise.

HAEMALOPIA (*he'-mul-o'-pe-ah*). A disease of the eye in which every object appears of a blood color.

HAEMOPHTHALMIA (*he'-mof-thal'-me-ah*). An effusion of blood into the eye.

HAEMOPHTHALMUS. Same as haemophthalmia.

HALO VISION. A condition in which a light or colored ring appears to encircle anything, especially a flame.

HARD CATARACT. A cataract having a hard nucleus.

HARLAN'S TEST. A test to detect malingerers.

HASNER'S VALVE. A fold in the nasal duct at its lower orifice.

HEBETUDO VISUS. Asthenopia.

HELOSIS. The eversion or turning out of the lids; also applied to convulsions of the muscles of the eye.

HEMERALOPIA (*hem'-ur-a-loh'-pe-ah*). A defect of vision by which objects are seen only in broad daylight; day-sight; night-blindness. It is also applied to a disorder of vision in which objects cannot be seen well, or without pain, by daylight.

HEMIACHROMATOPSIA (*a-kroh'-ma-top'-se-ah*). Obliteration of the color sense in one-half of the visual field.

HEMIANOPSIA (*hemi'-an-op'-se-ah*). Obscuration of one-half of the visual field.

HEMIOPALGIA (*hemi'-o-pal'-ge-ah*). Hemicranic pain of the eye.

HEMIOPIA (*oh'-pe-ah*) OR HEMIOPSIS. Disordered vision in which a patient sees only the half of an object.

HEN-BLINDNESS. Inability to see except by daylight; so termed because hens are said to be subject to it.

HENLE'S (*hen'-leez*) GLANDS. Tubular glands in the conjunctiva.

HERPES (*hur'-peez*) CORNEAE. A vesicular eruption on the cornea.

HETERONYMOUS (*het'-ur-on-i-mus*) DIPLOPIA. Same as crossed diplopia. In which the image of the left eye is on the right side, in which case the visual axes are divergent.

HETEROPHORIA (*fo'-re-ah*). A condition in which one of the visual axes tends to deviate from the point of fixation.

HETEROPHTHALMUS. The condition in which the color of one iris is different from the other.

HIPPUS. A pathological condition which consists in a constant and rapid change in the diameter of the pupil.

HOLMGREN'S TESTS. A set of colored skeins for the detection of color blindness.

HOMATROPINE (*hoh-mat'ro-peen*). An alkaloid derived from atropine and used as a mydriatic.

HOMER'S MUSCLE. Fibres of the ligamentum canthi internum which are inserted in the inner wall of the orbit.

HOMOCENTRIC (*hoh'mo-sen'trik*) RAYS. Rays which all either intersect at the same point or are parallel (i. e., intersect at infinity).

HOMONYMOUS DIPLOPIA (*hoh-mon'i-mus di-ploh'pe-ah*) A deviation of the axis of vision inward, thus projecting the image outward.

HORDEOLUM (*hawr-dee'o-lum*). A small furuncle in the margin of the lid.

HORIZONTAL MERIDIAN. A line drawn around the globe of the eye at right angles to the vertical diameter.

HORN OF THE EYELID. A cutaneous growth on the eyelid.

HOROPTER (*ho-rop'tur*). A line or surface in the field of vision, of such a shape that each point of it throws images upon corresponding points of the retinæ of the two eyes, and is hence seen as one point by both. It varies in character with the position of the eyes.

HYALITIS (*hy'a-ly'tis*). Inflammation of the vitreous humor.

HYALODECRYSIS (*hy'a-lo-dek'rih-sis*). Escape of part of the vitreous humor of the eye.

HYALODEOMALACIA (*hy'a-lo'de-o-may-lay'sih-ah*). Softening of the vitreous.

HYALOID ARTERY. An artery that traverses the vitreous anteroposteriorly in fœtal life.

HYALOIDITIS (*dey'tis* or *dee'tis*). Inflammation of the hyaloid membrane.

HYALOID MEMBRANE. The extremely delicate membrane which contains the vitreous humor.

HYDROPHTHALMUS (*hy-drof-thal'mus*). An enlargement of the eye in all directions.

HYOSCINE (*hy'o-seen*). An alkaloid derived from Hyoscyamus Niger.

HYOSCYAMINE (*hy'o-sy'a-meen*). An alkaloid derived from Hyoscyamus Niger; more soluble than Hyoscine; used as a mydriatic.

HYPÆMIA (*hip-e'me-ah*). A collection of blood in the anterior chamber.

HYPERBOLIC LENSES. Lenses for correcting the error of conical cornea.

HYPERESOPHORIA (*es'oh-foh'ree-ah*). A tendency of the visual lines upward and inward.

HYPEREXOPHORIA (*ek'soh-foh'ree-ah*). A tending of the visual lines upward and outward.

HYPERMETROPIA. That condition of the eye in which, when the accommodation is relaxed, the rays are brought to a focus behind the retina.

HYPEROPIA (*hy'pur-oh'pee-ah*). The same as hypermetropia.

HYPEROPES (*hy'pur-ope*). Those persons who are affected with hyperopia.

HYPEROPSIA (*hy'pur-op'se-ah*). Extremely acute vision.

HYPERPHORIA (*foh're-ah*). A tending of one visual line in a direction above its fellow.

HYPERTONIA (*loh'ne-ah*). Excessive intraocular tension of the globe.

HYPOCHYMA (*hy-pok'e-mah*). An old name for cataract.

HYPOGALA (*hyp'og-a-lah*). The effusion of a milk-like fluid into the chambers of the eye.

HYPOMETROPIA (*hy'po-me-tro'pe-ah*). A scientific name for the condition usually known as myopia.

HYPOPYON (*hy-poh'pe-un*). A collection of pus in the anterior chamber.

HYPOSPHAGMA (*hy'pos-fag'mah*). Subconjunctival hemorrhage.

HYPOTONIA (*loh'ne-ah*). Diminished tension of the globe.

HYSTERICAL ASTHENOPIA. Weakness of vision often occurring in hysterical patients.

ICE-BLINDNESS. A disease of the eyes caused by the brilliant reflection of the sun from the ice.

IDIOPATHIC (*id'e-oh-path'ik*) **IRITIS.** A form of iritis in which no local injury or constitutional disease can be accredited to its origin.

IDIOPATHIC RETINITIS. Retinitis occurring without known cause.

ILLACRYMATIO (*il-lak'-re-ma'-she-o*). Excessive involuntary weeping; sometimes synonomous with Epiphora.

ILLAQUEATION (*il-lak'we-a'shun*). An old operation for changing the position of an eye-lash by encircling its base in a loop made by a thread passed through the tissues.

IMAGE. The spectrum or picture of an object formed by the reflection or refraction of rays of light from its various points.

INCARCERATION (*in-kahr'-sur-a-shun*) **OF THE IRIS.** A condition in which a portion of the iris becomes adherent to a corneal scar.

INCIDENT RAY. The name given to a ray of light before its passage into the second medium.

INCIPIENT (*in-sip'-e-ent*) **CATARACT.** The first stage of cataract.

INDEX OF REFRACTION. The refractive power of any substance compared with that of water.

INFERIOR OBLIQUE (*ob-like', ob-leek'*) **MUSCLE.** The ocular muscle that rotates the cornea upward and outward.

INFERIOR RECTUS MUSCLE. The ocular muscle that rotates the globe downward.

INSUFFICIENCY OF THE OCULAR MUSCLES. Weakness of the muscles of one or both eyes, causing inability to move the globe so as to secure binocular vision.

INTERMITTENT STRABISMUS (*stra-biz-mus*). Strabismus which appears and disappears suddenly.

INTERNAL RECTUS MUSCLE. An ocular muscle, the function of which is to rotate the globe inward.

INTERPALPEBRAL SPOT. The triangular yellowish patch bordering the cornea on either side in old people; due to hypertrophy and colloid infiltration of the conjunctiva as a result of irritation arising from dust, etc. Same as pinguecula.

INTERPUPILLARY DISTANCE. The distance between the centres of the pupils.

INTERSTITIAL KERATITIS (*ker'-a-ty'-tis, tee'-tis*). A diffuse inflammation of the whole thickness of the cornea.

INTRA-OCULAR TENSION. A term used to designate the degree of pressure of the fluids of the eye.

IRALGIA (*i-ral'-gih-ah*). Pain of the iris.

IRIANKISTRIUM (*ir-ih-an-kis'-trih-um*) or IRIANKISTRON. A hook-shaped instrument used in the operation for artificial pupil by separation.

IRIDAEMIA (*ir-ih-de'-me-ah*). Hemorrhage from the iris.

IRIDALGIA (*ir-id-al'-jah*). Pain of the iris. Iralgia.

IRIDAUXESIS (*ir'-i-doh-e'-sis*). Thickening or growth of the iris by the exudation of fibrin into its substance.

IRIDECTOMUS (*ir'-i-dek'-to-mus*). A kind of knife used for iridectomy.

IRIDECTOMY (*ir'-i-dek'-to-mee*). The operation of removing or cutting out a portion of the iris.

IRIDECTROPIUM. Eversion of a portion of the iris.

IRIDENTROPIUM. Inversion of a portion of the iris.

IRIDENCLEISIS. An operation for displacing the pupil from its natural position, effected by drawing the iris into a wound made near the periphery of the cornea and causing it to become adherent there.

IRIDEREMIA (*ee-ree'-mee-ah*). Defect or imperfect condition of the iris.

IRIDES. Plural of iris.

IRIDESCENT VISION. The condition in which variously hued borders surround artificial lights.

IRIDESIS (*i-rid'e-sis*). Strangulation of a part of the iris to form an artificial pupil.

IRIDO-AVULSION (*ir'-i-doh-a-vul'-shun*). A term applied to the total removal of the iris by tearing it away from its periphery.

IRIDOCELE (*seel*). Hernia of the iris.

IRIDO-CHOROIDITIS (*koh'-ree-oy-dy'-tis, dee'-tis*). Inflammation of the iris and choroid coat of the eye.

IRIDOCINESIS (*ir'-i-doh-sih-ne'-sis*). The movement of the iris, its contraction and expansion.

IRIDOCYCLITIS (*ir'-i-doh-si-kli'-tis*). Inflammation of the iris and ciliary body.

IRIDODIALYSIS (*ir'-i-doh-di-al'-i-sis*). The operation for artificial pupil by separation.

IRIDODONESIS (*ir'-i-doh-nee'-sis*). Trembling of the iris.

IRIDOMALACIA (*ir'-i-doh-ma-la'-she-a*). Softening of the iris.

IRIDONCUS (*ir'-i-don'-kus*). Tumor or swelling of the iris.

IRIDOPERIPHACITIS (*pur'-i-fa-si'-tis*). Inflammation of the capsule of the lens of the eye.

IRIDOPLANIA. Trembling of the iris; iridodonesis.

IRIDOPLEGIA (*plee'-jah*). Paralysis of the iris.

IRIDORHEXIS. Tearing away of the margin of the iris.

IRIDORRHAGAS (*dor'-rha-gas*) Fissure of the iris.

IRIDOTOMY (*dot'-o-mee*). The operation for artificial pupil by incision.

IRIDOTROMUS (*dot'-ro-mus*). Trembling of the iris.

IRIS (*cy'ris*). The pigmented membrane separating the anterior and posterior chambers of the eye.

It is pierced by a central circular hole (the pupil).

It consists of the circular muscular fibres (sphincter of the iris) by which the pupil is contracted, of radiating elastic fibres by which the pupil is dilated, and of a posterior pigment layer which really belongs to the retina.

The iris is attached to the sclero-cornea by the ligamentum pectinatum.

IRITICUS. Iritic; belonging to iritis.

IRITIS (*cy'-ri-tis, ree'-tis*). Inflammation of the iris.

IROTOMY. The same as iridotomy.

IRREGULAR ASTIGMATISM. The condition when the refraction of the eye no longer presents any uniformity.

ISCHAEMIA (*is-ke'-me-ah*) RETINAE. Diminution of arteries in the retina.

ISCHURIOPHTHALMIA (*is-kew'-ry*). Ophthalmia resulting from the suppression of urine.

ISOMETROPIA (*cy'-so-me-tro-pe'-ah*). The state in which both eyes are alike in their refraction.

JABORANDI (*jab'ur-an'dee*). The Pilocarpus pennatifolius, a South American shrub of the Rutaceæ. The leaflets produce marked sweating, salivation, increase of the milk and other secretions, miosis and spasm of the accommodation, lowering of the blood pressure and temperature and often marked prostration. These effects are due to the presence of an alkaloid, pilocarpine. They also contain the alkaloid, jaborine, which acts like atropine. J. and pilocarpine are used as diaphoretics in detachment of the retina; as a miotic and to reduce the intra-ocular tension in glaucoma, staphyloma and certain ulcers of the cornea.

JEQUIRITY (*je-kwir'i-tee*). An Indian plant, the infusion of which is sometimes used as a remedy against chronic granular ophthalmia. It is very irritating to the eyes and should not be used.

KERATALGIA (*ker'at-al'jah*). Neuralgic pain of the cornea.

KERATITIS (*ker'a-tv'tis, tee'tis*). Inflammation of the cornea.

KERATITIS BULLOSA (*bul-o'sah*). A disease in which small blebs form on the cornea.

KERATOCELE (*ker'a-toh-seel*). Hernia of the cornea.

KERATOCONUS (*koh'nus*). Conical cornea.

KERATOGLOBUS (*tog'lo-bus*). A globular protrusion of the cornea.

KERATOHELCOSIS (*hel-koh'sis*). Ulceration of the cornea.

KERATO-IRITIS. Inflammation of both cornea and iris.

KERATOKINESIS (*ky-ne'sis*). The formation of new cells in the cornea.

KERATOMALACIA (*ma-lay'shah*). A purulent infiltration of the whole cornea. It is very destructive.

KERATOMETER (*tom'ut-ter*). An instrument for measuring the cornea.

KERATONYXIS (*nik'sis*). An operation for cataract in which the lens is depressed by a needle passed through the cornea. Paracentesis or any puncture of the cornea.

KERATOPLASTIC. Belonging to keratoplasty.

KERATOPLASTY (*ker'a-to-plas'tee*). The repair by operation of defects or redundancies of the cornea; especially, the substitution by operation of transparent for opaque cornea.

KERATOSCOPE (*ker'a-toh-skope*). An instrument for examining the cornea; especially, one for determining from inspection of the cornea the form and curvature of the latter.

KERATOSCOPY (*ker'a-tos'ko-pee*). Examination of the cornea with a keratoscope. Skiascopy.

KERATOTOME (*ker'a-to-tome*) or KERATOME. A knife for incising the cornea.

KERECTOMY (*ker-ek'to-me*). The operation for obtaining a clear space in an opaque cornea.

KOPIOPIA or COPIOPIA (*kop'ee-oh'pee-ah*). Asthenopia.

KORECTOMIA or CORECTOMIA (*ko'rek-to'mee-ah*). The operation for artificial pupil by removal of a part of the iris.

KOROSCOPY (*ko-ros'ko-pee*). A method of determining the refractive state of the eye by examining the movement of light and shadow across the pupil when the retina is illuminated by light thrown into the eye from a moving mirror.

KRAUSE'S (*krow'zez*) GLANDS. Mucous glands in the fornix conjunctivæ.

LACHRYMAL (*lak'-ri-mul*). Of or pertaining to tears.

LACHRYMAL ABSCESS. An abscess of the lachrymal sac.

LACHRYMAL APPARATUS. Consists of the lachrymal gland, lachrymal ducts, canaliculi, sac and nasal ducts.

LACHRYMAL GLAND. The gland situated at the upper and outer angle of the orbit. This gland secretes the tears.

LACHRYMATION (*lak'-ri-may'-shun*). An increased flow of tears.

LACUNAR ORBITAE (*la-kew'-nur or'-bit-a*). The roof of the orbit of the eye.

LAGOPHTHALMUS (*lag'-of-thal'-mus*). Complete absence of the eyelids.

LAMINA CRIBROSA. A fine web of fibrous tissue on the posterior surface of the sclera for the passage of the optic nerve.

LAMINA FUSCA. The external layer of the choroid.

LAPIS (*lay'-pis*) DIVINUS. An application for follicular granulations, made of equal parts of sulphate of copper, nitrate of potash, and alum, fused together and moulded.

LATENT HYPEROPIA. That part of the hyperopia which the crystalline lens can overcome.

LEAD OPACITY. A corneal opacity caused by the precipitation of lead salts from washes containing lead.

LENS. A glass which, owing to its peculiar form, causes the rays of light to converge to a focus or disperses them according to the laws of refraction.

Crystalline l. A transparent, double convex lens situated between the aqueous and vitreous humors of the eye.

Convex l. A lens which converges the rays of light.

Concave l. A lens which disperses the rays of light.

Cylindrical l. A lens formed from the segment of a cylinder, and may be either convex or concave.

Compound l. A lens consisting of several lenses put together.

LENTICONUS (*len'-tee-koh'-nus*). A transparent conical projection from the posterior surface of the lens.

LENTICULAR (*len-tik'-yu-lur*) FOSSA. The depression in the anterior part of the vitreous for the lens.

LENTICULAR GANGLION. A small reddish body near the back part of the orbit between the optic nerve and the external rectus muscle.

LEPROPHTHALMIA (*lep'-rof-thal-me-ah*). Leprous ophthalmia.

LEPTOTHRIX (*lep'-to-thriks*). A mass of fungus in one of the canaliculi.

LEUCOMA (*lew-ko'-mah*). Opaque cicatricial tissue in the cornea.

LEVATOR PALPEBRAE MUSCLE. The muscle which raises the upper lid.

LIDS. The anterior protective coverings of the eye.

LIGAMENTUM PECTINATUM (*pek'-ti-na'-tum*). That part of Descemet's membrane which is reflected on the iris.

LIGHT. Light is that great force of nature by the action of which objects are made visible.

LIGHT SENSE. The power possessed by the retina of appreciating variations in the intensity of the source of illumination.

LIMBUS CORNEAE. The edge of the cornea.

LINE OF FIXATION. The line which connects the object looked at, with the center of rotation of the eye.

LINE OF VISION. The line which connects the object looked at with the fovea centralis.

LIPPITUDO (*lip'-pee-tew-doh*). The appearance caused by the exposure of the marginal conjunctiva together with the loss of the cilia.

LIQUOR MORGAGNI. A small quantity of fluid between the lens and its capsule.

LOIMOPHTHALMIA. Contagious ophthalmia.

LONG SIGHTEDNESS. Presbyopia

LOUCHETTES. A kind of opaque glasses in which for each eye there is a small hole, which makes it impossible to look in any other way than through this opening.

LOXOPATHALMUS (*lok-sop'-thal-mus*). Having oblique or squinting eyes.

LUXATION (*luk-sah'-shun*) OF LENS. Dislocation of the crystalline lens.

MACROPSIA. An affection of the eye in which objects appear larger than they really are.

MACULA LUTEA (*mak'-yu-lah lew'-te-ah*). The yellow spot; a depressed elliptical or circular spot at the centre of the retina; the point of the most acute vision. It contains a central depression, the fovea centralis.

MADAROSIS (*mad'-ur-oh'-sis*). The condition in which the lashes are permanently destroyed.

MALINGERING. Pretending blindness to escape irksome duty or to excite sympathy.

MANIFEST HYPEROPIA. The hyperopia remaining after the crystalline lens has exerted all of its power.

MARGINAL KERATITIS. A form of keratitis characterized by the development of numerous phlyctenules around the rim of the cornea.

MARMARYGAE (*mar-mar'-ig-e*). The appearance of sparks or coruscations before the eyes.

MEGALOPSIA (*meg'-a-lohp'-se-ah*). An affection of the eye in which objects appear larger than they really are.

MEGOPHTHALMUS (*of-thal'-mus*). A condition in which the whole eye is abnormally large.

MEIBOMIAN (*mey-boh'-mee-un*) GLANDS. Small glands between the conjunctiva and the tarsal cartilages. When acutely inflamed they produce a Meibomian sty (Hordeolum meibomianum); when enlarged by obstruction of the duct and thickening of the walls they form a chalazion.

MELANOPHTHALMIA (*mel'-an*). A melanotic tumor of the eye; it is characterized by the deposition of black pigment.

MELANOSIS SCLERAE. Congenital pigmentation of the sclera.

MEMBRANA LIMITANS INTERNA. The internal layer of the retina.

MEMBRANA NICTITANS. A thin membrane forming a third eyelid in certain kinds of birds.

MEMBRANA PUPILLARIS. A membrane covering the pupil in fœtal life. This sometimes fails to disappear.

MEMBRANOUS OPHTHALMIA. Same as diphtheritic ophthalmia.

MENISCUS GLASSES. Glasses that refract at some distance from the centre, the same as at the centre, so that persons can see obliquely through them.

MEROPIA (*me-roh'-pe-ah*). Partial dulness or obscuration of sight.

METAMORPHOPSIA (*matwr-fop'-see-ah*). An affection of the eyes in which objects appear changed from their natural form.

METRE ANGLE. The angle through which the visual line must move upon parallelism to fix an object one metre distant.

METRE LENS. A lens, the focal length of which is one metre.

MICROPHTHALMIA (*mey'-krof-thal'-mee-ah*). A morbid shrinking or wasting of the eyeball.

MICROPSIA (*mey-krop'-see-ah*). An affection of the eye in which objects appear smaller than they really are.

MILITARY OPHTHALMIA. A purulent contagious ophthalmia, same as Egyptian ophthalmia.

MILIUM. A small elevation filled with sebum on the skin of the eyelid.

MITIGATED STICK. Nitrate of silver, one part, and nitrate of potash, two parts, fused together and moulded into sticks.

MOLLUSCUM CONTAGIOSUM. A disease of the sebaceous glands, characterized by the appearance of rounded papules, about the size of a pea and of waxy color.

MONOBLEPSIS. A state of vision in which objects are distinct only when viewed with one eye.

MOON-BLINDNESS. Functional night blindness.

MORGAGNIAN (*mor-gag'-ni-an*) **CATARACT.** A type of cataract in which the cortical matter has liquefied and the nucleus has been displaced.

MUCOCELE (*mew'ko-seel*). A distention of the lachrymal sac with clear or turbid mucus

MUSCAE VOLITANTES (*mus'ee vol'i-tan'teez*). Black specks seen floating in the field of vision. A sympton due to opacities in the media, especially in the vitreous humor.

MUSCARINE (*mus'kar-een*). An alkaloid obtained from the albumen of hens' eggs. It acts as a myotic and causes contraction of the ciliary muscle.

MUSCULAR ASTHENOPIA. Weakness of the ocular muscles.

MYCOPTHALMIA (*mey'kof-thal'mee-ah*). Fungous inflammation of the eye.

MYDRIASIS (*mid'rey'a-sis*). Dilatation of the pupil.

MYDRIATICS (*mid'ree-at-iks*). Drugs that cause dilatation of the pupil.

MYOCEPHALON (*mey'oh-seph'a-lon*). A knuckle of iris protruding into a corneal ulcer.

MYODESOPSIA (*mey'oh-de-sop'see-ah*). Same as muscae volitantes.

MYOPIA (*mey-oh'pee-ah*). The condition of the refraction when the retina is behind the focus.

MYOPIC CRESCENT (*mey-op'ik kres'ent*). A crescent shaped atrophy of the choroid at the posterior pole of the eye.

MYOSIS (*mey-oh'sis*). Contraction of the pupil.

MYOTICS (*mey-ot'iks*). Drugs that cause contraction of the pupil.

NASAL DUCT. That part of the tear duct below the lachrymal sac, and opening into the nose.

NEAR POINT. The nearest point at which the eye still has maximum visual acuity.

NEAR SIGHTEDNESS. Myopia.

NEBULA CORNEAE (*neb'yu-lah kaur'nee-ah*). A superficial cloudy condition of the cornea from loss of the epithelium.

NECROSIS CORNEAE. A disease characterized by dryness of the conjunctiva, and destructive ulceration of the cornea.

NEPHRITIC (*nee-frit'ik*) **RETINITIS.** A form of inflammation of the retina associated with Bright's disease of the kidneys.

NEURODEALGIA. Pain or excessive sensibility of the retina.

NEURODEATROPHIA. Atrophy of the retina.

NEURO-PARALYTIC KERATITIS. An anaesthetic condition of the cornea coupled with ulcerative inflammation.

NEURO-RETINITIS. Inflammation of the optic nerve and retina.

NICTITATION. Involuntary convulsive twitching of the eyelids.

NIGHT-BLINDNESS. Hemeralopia.

NIPHABLEPSIA (*nif-ah-blep'se-ah*). Blindness caused by the glaring reflection of sunlight upon the snow.

NIPHOTYPHLOTES (*nif-o-tif-lo'tez*). Same as niphablepsia.

NITRATE OF SILVER. A drug, the solution of which is often used in purulent ophthalmias and as a stimulant application in chronic trachoma.

NODAL POINTS. These are two points in a lens, the incident ray being directed toward the first and the refracted ray toward the second. The optical centre lies between them.

NUCLEAR (*new'klee-ur*) CATARACT. A form of cataract in which the nucleus of the lens is opaque.

NYCTALOPIA (*nik'ta-loh'pee-ah*). Day-blindness.

NYCTOTYPHLOSIS (*nik'to-tey-flo'sis*). A term for night-blindness.

NYSTAGMUS (*nis-tag'mus*). A term applied to a condition characterized by an involuntary, rapid movement of the eyeballs, either from side to side, vertically, or in a rotary direction.

OBFUSCATION (*ob'fus-ka'shun*). Obscure sight. A clouding; as O. of the cornea.

OCCLUSION (*ok-kloo'zhun*) OF THE PUPIL. Blocking up of the pupil by a membrane.

OCULAR. Belonging to the eye.

OCULAR CONE. A cone formed in the eye by the rays of light, the base being on the cornea, and the apex on the retina.

OCULAR SPECTRES. Imaginary objects floating before the eyes.

OCULI (*ok'yu-ley*). Plural of oculus.

OCULIST (*ok'yu-list*). One skilled in diseases of the eye.

OCULOMOTOR (*ok'yu-loh-moh'tur*). Of or pertaining to the movement of the eye; as the O. nerve (oculomotorius) the third cerebral nerve which innervates all the muscles of the eye except the superior oblique and the external rectus.

OCULO-NASAL. Pertaining to or supplying the eye and nose.

OCULUS (*ok'yu-lus*). The organ of vision.

O. D. Abbreviation for oculus dexter (right eye).

OLD SIGHT. See presbyopia.

ONYX (*on'iks*). An accumulation of pus between the layers of the cornea.

OPHTHALMAGRA (*of-thal'ma-gra*). Sudden pain in the eye, usually gouty in origin.

OPHTHALMALGIA (*of-thal-mal'ge-ah*). Sudden violent pain in the eye, not the result of inflammation.

OPHTHALMALGICUS (*of-thal-mal'ge-kus*). Belonging to ophthalmalgia.

OPHTHALMATROPHIA (*of-thal-mah-tro'fe-ah*). Atrophy of the eye.

OPHTHALMIA (*of-thal'mee-ah*). Inflammation of the eye. Catarrhal o., the severer forms.

　　Egyptian o., trachoma. Gonorrhœal o., Purulent o., acute blennorrhœa of the conjunctiva; gonorrhœal conjunctivitis.

Jequirity o., purulent conjunctivitis produced by the instillation of an infusion of jequirity into the eye.

Metastatic o., chorioiditis due to pyaemia or other form of metastatic infection. Neuro-paralytic o., keratitis neuro-paralytica. O. neonatorum. A specific purulent ophthalmia of infants. Phlyctenular o., phlyctenular conjunctivitis and keratitis. Sympathetic o., a destructive, usually recurrent, plastic irido-cyclitis occurring in one eye as a result of injury or inflammation of its fellow.

OPHTHALMIC (*of-thal'mic*). Belonging to or connected with the eye, or with ophthalmia.

OPHTHALMITIC. Belonging to ophthalmitis.

OPHTHALMITIS (*mey'tis, mee'tis*). Inflammation of the eye, more especially the globe with its membranes.

OPHTHALMOBLENNORRHŒA (*blen-ur-ree-ah*). A flow of mucus from the eye.

OPHTHALMOCARCINOMA (*kahr-si-noh'mah*). Cancer of the eye.

OPHTHALMOCELE. Protrusion of the eyeballs.

OPHTHALMOCELICUS. Belonging to ophthalmocele.

OPHTHALMOCOPIA (*koh'pee-ah*). Asthenopia.

OPHTHALMODYNAMOMETER. An instrument to determine the maximum of convergence.

OPHTHALMODYNIA (*din'e-ah*). Sudden violent pain in the eye not the result of inflammation.

OPHTHALMOGRAPHY (*mog'rha-fee*). A description of the eye.

OPHTHALMOLOGY (*mol'-o-jee*). A treatise on the eye.

OPHTHALMOMACROSIS (*ma-kro'sis*). Enlargement of the eyeballs.

OPHTHALMOMALACIA (*ma-la'sha*). A condition in which, without known cause, the eyeball shrinks and becomes soft, but returns after a time to its normal state.

OPHTHALMOMETER (*mom'e-ter*). An instrument for measuring the eye; particularly, an instrument for determining the amount of astigmatism by examination of images reflected from the surface of the cornea.

OPHTHALMOMETRY (*mom'et-ree*). The determination of the refraction by the ophthalmometer.

OPHTHALMOPATHY (*mop'a-thee*). Any affection of the eyes.

OPHTHALMOPHTHISIS (*mof'thi-sis*). Wasting of the eyeballs. Phthisis Bulbi.

OPHTHALMOPLEGIA (*ple'jah*). Paralysis of the muscles of the eye.

OPHTHALMOPTOMA (*mop-to'mah*). Protrusion of the eyeballs.

OPHTHALMOPTOSIS (*mop'to-sis*). The progress of ophthalmoptoma.

OPHTHALMORRHAGIA (*mor-rha'gee-ah*). Hemorrhage from the eye or orbit.

OPHTHALMORRHEXIS (*mor-rex'is*). Bursting of the eyeball.

OPHTHALMORRHŒA (*mor-ree'ah*). An oozing of blood from the eye.

OPHTHALMORRHŒA EXTERNA. Extravasation of blood beneath the eyelids.

OPHTHALMORRHŒA INTERNA. Extravasation of blood within the eye.

OPHTHALMOSCOPE (*of-thal-mo-skohp*). An instrument, consisting usually of a perforated mirror, for examining the interior of the eye, and thus determining the appearance of the media, the condition of the retina, choroid, and optic nerve, and the state of the refraction.

The light is reflected by the mirror into the eye, is reflected thence, passes through the hole in the mirror, and enters the eye of the observer, which is placed behind the o.

In the direct method of using the o. the latter is held close to the eye examined and an erect virtual image of the fundus is obtained.

In the indirect method an inverted real image of the fundus is formed in front of the patient's eye by means of an auxiliary lens held before the latter, and this image is then examined by the observer, who stations himself some distance from the patient.

OPHTHALMOSCOPIC (*mos-kop'ik*). Belonging to ophthalmoscopy.

OPHTHALMOSCOPY (*mos'co-pee*). The use of the ophthalmoscope.

OPHTHALMOSTAT (*of-thal'mos-tat*). An eye-speculum.

OPHTHALMOSTATOMETER (*sta-tom'-e-ter*). An apparatus for determining the degree of prominence of the eyeball.

OPHTHALMOTONOMETER (*toh-nom'e-ter*). An instrument for measuring the tension of the eyeball.

OPHTHALMOTONOMETRY (*toh-nom'ut-ree*). The determination of the intra-ocular tension.

OPHTHALMOTROPE (*of-thal'mo-trope*). An artificial eye which can be made to rotate about its center and imitate the movements of the natural eye.

OPHTHALMOTROPOMETER (*troh-pom'e-ter*). An apparatus for measuring the movements of the eye.

OPTICAL. Relating to the organ of vision.

OPTIC AXIS. An imaginary line passing through the centre of the cornea and lens and the point of rotation, to the posterior pole of the eye.

OPTIC NERVE. The nerve which forms the communication between the organ of vision and the brain.

OPTIC DISC. The flat terminal expansion of the optic nerve upon the retina.

OPTIC PAPILLA (*pa-pill'ah*). Same as optic disc.

OPTICS. That branch of physical science which treats of light and vision, and the instruments by the use of which the faculty of vision is aided and improved.

OPTIC THALAMUS (*thal'a-mus*). Each of two eminences in the anterior and internal part of the lateral ventricles of the brain. The bed of the optic nerve.

OPTIC TRACT. The course of the optic nerve before it reaches the commissure.

OPTOMYOMETER (*mey-om'e-ter*). An instrument for measuring the power of the ocular muscles without exciting accomodation.

ORA SERRATA (*oh'ra seh ray'tah*). The anterior limit of the retina.

ORBICULARIS (*aor-bik'yu-lah'-ris*) OCULI. The same as orbicularis palpebrarum.

ORBICULARIS PALPEBRARUM. The circular muscle of the eyelids.
A muscle arising from the outer edge of the orbital process and inserted into the nasal process of the superior macillary bone. It shuts the eye.

ORBIT. The bony cavity in which the eyeball is situated.

ORBITAL. Belonging to the orbit.

ORBITARY. Relating to the orbit of the eye.

ORTHOPHORIA (*foh'ree-ah*). The condition in which the eyes are properly placed with respect to each other.

ORTHOSCOPE (*aur-tho-scope*). An instrument for neutralizing the refraction of the cornea.

ORTHOSCOPIC SPECTACLE GLASSES. Two spectacle glasses cut out of a large one in such a way that each eye may be furnished with that portion of the lens which is on the eye's axis when the lens is whole.

O. S. Abbreviation for oculus sinister (left eye).

PACHYBLEPHARUM (*pak'ee-blef'ur-um*). A thickening of the eyelid, from obstruction of the Meibomian glands.

PACHYBLEPHAROSIS (*blef-ur-oh'sis*). The progress of pachyblepharum.

PAGENSTECHER'S (*Pah'gen-stech-erz*) OINTMENT—
yellow oxide of mercury . . . . gr. iv to vii.
vaseline . . . . oz. i.

PANNUS. The presence of bloodvessels in the cornea; a kind of vascular keratitis.

PANOPHTHALMITIS (*pan-of'thal-mey'tis, mee'tis*). Inflammation of the entire eye, ending in rupture and total destruction.

PAPILLITIS (*pap-il-ey'tis, ee'tis*). Inflammation of the optic papilla,

PAPILLO-RETINITIS. Inflammation of the papilla and retina.

PARABLEPSIS. False vision.

PARACENTESIS (sen-tee'sis) CORNEA. Puncture of the cornea.

PARENCHYMATOUS (en-kim'-a-tus) IRITIS. A form of iritis characterized by discoloration and swelling of the iris caused by cellular proliferation within its tissues.

PAROPSIS (par-op'-sis). A generic term for disorder of vision.

PERICHOROIDAL (koh'ree-oy'-dul). Surrounding the choroid membrane.

PERIMETER (per-im'e-tur). A instrument for examining the peripheral parts of the retina.

PERISCOPIC GLASSES. Concavo-convex and convexo-concave lenses.

PERITOMY (per-it'o-mee). Section of the conjunctiva around the cornea.

PHACO (fak'oh). Prefix meaning of or pertaining to a lens, especially the crystalline lens.

PHACOCYSTA (fak'o-sis'tah). The capsule of the crystalline lens.

PHACOCYSTECTOME (sis-tek'to-me). Rognetta's operation for cataract by cutting out a part of the capsule.

PHACOCYSTITIS (sis-tey'tis, tee'tis). Inflammation of the capsule of the lens.

PHACOHYMENITIS (hey-men-cy'tis, ee'tis). Inflammation of the capsule of the lens.

PHACOMALACIA (ma-lay'she-ah). Softening of the crystalline lens.

PHACOMETER. An instrument for measuring the curvature of lenses, and so determining their refractive power and, if cylindrical, their axis.

PHACOSCLEROSIS (skle-roh'sis). Sclerosis of the crystalline lens. The process which produces hard cataract.

PHACOSCOPE (fak'o-skope). An instrument for examining the images reflected from the anterior and posterior surfaces of the crystalline lens, and thus determining the changes which the latter undergoes in accommodation.

PHAKITIS (fa-key'tis). Inflammation of the lens capsule.

PHANTASMA. A disease of the eye in which imaginary objects are seen.

PHENGOPHOBIA. A fear or intolerance of light.

PHOSPHENES (fos'feenz). Subjective phenomena of light caused by external mechanical means.

PHOTALGIA (fo-tal'ge-ah). Pain arising from too much light.

PHOTOCAMPSIS (fo-toh-kamp'sis). Refraction of the rays of light.

PHOTODYSPHORIA (dis-fo'-re-ah). Intolerance of light.

PHOTOLOGY. The science of light.

PHOTOMETER (tom'e-tur). An instrument for testing the light sense.

PHOTONOSUS (*fo-ton'-o-sus*). Any disease of the eye arising from exposure to a glare of light.

PHOTOPHOBIA (*foh'bee-ah*) Intolerance of light.

PHOTOPSIA (*top-see'ah*). Phosphenes; flashes of light; luminous rings, etc.

PHTHIRIASIS (*thur-ee-ey'sis*). Blepharitis pediculosa.

PHTHISIS (*tey'sis, tee'sis*) BULBI. Shrinking of the eyeball.

PHYSIOLOGICAL CUP. The normal cup-shaped depression at the entrance at the head of the optic nerve.

PHYSOSTIGMINE (*fey'soh-stig'meen*). An alkaloid derived from Calabar bean; it is used as a myotic.

PILOCARPINE (*pey'loh-kahr'peen*). An alkaloid derived from jaborandi; it is used as a myotic.

PINGUECULA (*ping-gwek'yu-lah*). A small yellowish elevation situated in the conjunctiva near the margin of the cornea.

PINK EYE. Catarrhal conjunctivitis.

PLICA (*ply'kah*) SEMILUNARIS. A fold of conjunctiva near the inner canthus.

POLYCORIA (*koh-ree'ah*). A multiplicity of pupils.

PRESBYOPIA (*prez'bee-oh'pee-ah*). The condition as a result of age and the consequent inability to accommodate.

PRINCIPAL MERIDIANS. The vertical and horizontal meridians of the cornea.

PRISM. A prism is a transparent portion of glass or transparent substance between two plane surfaces which are inclined to each other.

PRISM DIOPTRE (*dey-op'tur*). A prism which deflects a ray of light one centimetre at a plane one metre distant.

PROPTOSIS (*prop-toh'sis*). A protrusion of the eyeball.

PROTHESIS (*pro-the'sis*) OCULARIS. The insertion of an artificial eye.

PSEUDO GLIOMA (*sew'doh gly-oh'mah*). A circumscribed collection of pus in the vitreous.

PSOROPHPHALMIA (*soh'rof-thal'mee-ah*). Inflammation of the eye attended with itchy ulcerations.

PTERYGIUM (*tee-rij'ee-um*). A fan-shaped fleshy growth consisting of hypertrophy of the conjunctiva, extending from the inner angle of the eye to the cornea; it rarely passes the centre of the cornea.

PTOSIS (*toh'sis*). Drooping of the upper lid.

PTOSIS IRIDIS. A prolapse of the iris through a lesion or wound of the cornea.

PTOSIS LIPOMATOSIS (*lih-poh'ma-toh-sis*). An extensive accumulation of fat in the connective tissue of the upper lid causing it to droop.

PUNCTA LACHRYMALIA (*pung k'ta-lak-ri-may'lee-ah*). The openings of the canaliculi.

PUPIL (*pew'pil*). The central, circular opening in the iris.

PUPILLOMETER. An instrument for measuring the pupil.

PUPILLOSCOPY. Skiascopy.

QUININE AMBLYOPIA. Loss of vision from excessive use of quinine.

RED BLINDNESS. The inability to distinguish red.

REFRACTION. The deviation of a ray of light from its original direction on entering obliquely a medium of different density.

RETINA. The ocular expansion of the optic nerve.

RETINITIS. Inflammation of the retina.

RETINITIS ALBUMINURICA. That form of retinitis caused by albuminuria.

RETINOSCOPY. The shadow test of refraction.

RETROBULBAR NEURITIS. Inflammation of the optic nerve behind the globe of the eye.

RHEUMATIC IRITIS (*rey'tis, ree'tis*). Iritis caused by rheumatism.

SCLERA. The external coat of the eyeball.

SCLERITIS. Inflammation of the sclera.

SCLERONYXIS (*nik'sis*). Cutting through the retina.

SCLEROPHTHALMIA. An opacity of the cornea caused by the sclera lapping over it.

SCOTOMA (*sko-toh'mah*). A blind spot on the retina.

SHORTSIGHTEDNESS. Myopia.

SKIASCOPY (*skey-as'ko-pee*). Retinoscopy.

SNOW BLINDNESS. Blindness from over-stimulation of the retina by rays reflected from snow.

SQUINT. Same as strabismus.

STAPHYLOMA (*staf'il-oh'mah*). Bulging of the cornea.

STRABISMOMETER (*stra-biz-mom'e-ter*). An instrument for measuring the degree of strabismus.

STRABISMUS (*stra-biz'mus.*) Cross-eyes.

STYE. Hordeolum; a furuncular affection of the connective tissue of the lids.

SURSUMDUCTION (*sur'sum-duk'shun*). The power of uniting the two images of the candle seen through a prism with its base down before one eye.

SURSUMVERGENS (*sur'sum-vur'jenz*). It means tending upward as in vertical squint.

SYMBLEPHARON (*sim-blef'ur-on*). An adhesion between the mucus membrane of the ball and that of the lid.

SYNCHYSIS (*sing'kih-sis*). A mingling of the humors of the eye in consequence of the rupture of the internal membrane and capsule by a blow.

SYNDECTOMY (*sin-dek'to-me*). Removal of a ring of conjunctiva around the cornea for the cure of pannus.

SYNECHIA (*si-nee'kee-ah*) ANTERIOR. Adhesion of the iris to the cornea.

SYNECHIA POSTERIOR. Adhesion of the iris to the lens.

TAPETUM. A lustrous, greenish membrane seen in the eyes of many animals.

TARSAL CARTILAGE. The cartilages that give the lid its shape.

TARSITIS (*tahr-sey'tis, see'tis*). Inflammation of the tarsal cartilages.

TARSOPHYMA (*tahr-so-fy'mah*). A morbid growth or tumor of the tarsus.

TARSORRHAPHY (*tahr-sor'a-fee*). The uniting by suture of any wound of the eyelids near the tarsus.

TARSOTOMY. A cutting of the tarsus, or of the cartilage of the eyelid.

TARSUS. The thin cartilage toward the edge of each eyelid giving it firmness and shape.

TEARS. The lachrymal secretion.

TENONITIS. Inflammation of the capsule of Tenon.

TENOTOMY (*tee-not'o-mee*). Cutting the tendons of the ocular muscles.

TINEA TARSI. Blepharitis marginalis.

TOBACCO AMAUROSIS. Dimness of vision from excessive use of tobacco.

TONOMETER (*toh-nom'e-ter*). An instrument for measuring the tension of the eyeball.

TRACHOMA (*tra-koh'mah*). Granulations of the lids.

TRICHIASIS (*tri-key'a-sis*). A disease characterized by irregularity in the insertion and direction of the cilia.

TYLOSIS (*tey-loh'sis*). A name for the thickened, ulcerated condition of the lid margins after ulceration.

URAEMIC (*ewr-ee'mik*). AMAUROSIS. Dimness of vision occurring in connection with uraemia.

UVEA (*ew've-ah*). The choroid, ciliary body, and iris.

UVEITIS (*ew'vee-ey'tis, ee'tis*). Inflammation of the uveal tract.

VISION. The act of seeing.

VISUAL ANGLE. The angle formed at the eye by rays coming from opposite extremities of an object.

VITREOUS HUMOR. The gelatinous refracting medium occupying the larger part of the globe.

WALL-EYE. A condition (especially in horses) in which the iris is whitish.

XANTHELASMA (*zan'the-laz'mah*). A slightly-raised yellow patch on the skin of the lids.

XEROMA (*zer-oh'mah*). Atrophy of the conjunctiva.

XEROPHTHALMIA (*zer-of-thal'mee-ah*). An abnormal dryness of the conjunctiva.

YELLOW SPOT. The macula lutea.

ZONULA OF ZINN. The suspensory ligament of the lens.

ZONULAR CATARACT. A form of cataract in which the opacity is limited to a few layers of the lens next to the nucleus.

PRESCRIPTIONS

FORMULA No. 1.

R. acid boracic, grs, v.
    aqua destil., oz, i.
    Mix and filter.

To be instilled into the eye three or four times a day in acute
hyperaemia.

FORMULA No. II.

R. hydrastin, gr, ss.
    acid carbolic (pure), gtt, i.
    morphia sulph.,
    cocaine murias, aa grs, iv.
    glycerine, dr's, ii.
    aqua destil., dr's, vi.
    Mix and filter.

A few drops in the eye three or four times a day.
To be used in the different forms of conjunctivitis.

FORMULA No. III.

R. argenti nitras (cryst.), gr, i.
    aqua destil., oz, i.
    Mix.

To be applied to the edges of the lids in blepharitis marginalis.

FORMULA No. IV.

R. chloride zinc, gr, i.
    cocaine murias,
    morphia sulphas, aa gr's, iv.
    acid carbolic, gtts, v.
    glycerine, dr's, ii.
    aqua rosa, dr's, vi.
    Mix and filter.

To be applied to corneal ulcer once a day with a cotton holder.

FORMULA No. V.

R. hydrastin, gr. ss.
    acid carbolic (pure), gtts, ii.
    cocaine murias, gr's, viii.
    glycerine, dr's. ii.
    aqua hamamelis (dist'd), dr's, vi.
    Mix and filter.

A few drops in the eye when painful in phlyctenular ulcers.

FORMULA No. VI.

R. hydrargeri oxidum flav., gr's, iv.
    vaselini, oz. i.
    Mix thoroughly.

To be applied to the edges of the lids in blepharitis marginalis.

FORMULA No. VII.

R. carbolic acid, gtt, i.
    hydrastin, gr, ss.
    boracic acid, gr's, x.
    cocaine hydrochlorate, gr's, iv.
    glycerine, dr's, ii.
    aqua hamamelis, dr's, vi.
    Mix and filter.

A few drops in the eye three times a day in the purulent ophthalmias.

In all of the above prescriptions where carbolic acid is employed, it should be first added to the glycerine, and the remaining ingredients to the aqua destil., aqua rosa, or aqua hamamelis (whichever may be indicated), then the two solutions united and filtered.

After any of the above formulas has been prepared and bottled, it should be placed in boiling water, in order that the contents are rendered perfectly aseptic.

INDEX.

www.ingramcontent.com/pod-product-compliance
Lightning Source LLC
Chambersburg PA
CBHW021347210326
41599CB00011B/787